SUFFOLK'S GARDENS AND PARKS

D1536694

Suffolk's Gardens and Parks

Designed Landscapes from the Tudors to the Victorians

TOM WILLIAMSON

WIND*gather*
PRESS

Copyright © Centre for East Anglian Studies, University of East Anglia, 2000

Published by: Windgather Press, 31 Shrigley Road, Bollington, Macclesfield, Cheshire SK10 5RD, UK in association with the Centre for East Anglian Studies, University of East Anglia, Norwich, UK. Published with the aid of a grant from the Ann Ashard Webb bequest.

Distributed by: Central Books, 99 Wallis Road, London E9 5LN

British Library Cataloguing-in-Publication Data
A catalogue record for this book is available from the British Library

Library of Congress Cataloging-in-Publication applied for
Suffolk's Gardens and Parks: Designed Landscapes from the Tudors to the Victorians, by Tom Williamson

First published 2000

Typeset and originated by Carnegie Publishing Ltd, Chatsworth Road, Lancaster
Printed and bound by Alden Press, Oxford

ISBN 0 9538630 0 X *paperback*

Contents

List of illustrations

Figures

Plates

The plates are reproduced between pages 84 and 85.

Acknowledgements

I wrote this book, but it rests firmly on the work of others. In particular, much of the information assembled here was collected by two researchers of extraordinary ability – Anthea Taigel and Monica Place – without whom this volume would have been thin indeed. Suffolk County Council generously supported research into the county's gardens for many years, and my thanks go, in particular, to Peter Holborn and the various members of his team – especially Lynn Dicker and Rose Morford – who provided much help and advice. I am also grateful to the many other Suffolk people who have assisted and encouraged this foray from across the Waveney; in particular Norman Scarfe, John Blatchley and Edward Martin. Thanks are also due to the many landowners who allowed me access to their properties and their archives: the list is a long one, but particular thanks go to Peter Strutt, Lord Somerleyton, Lord and Lady Henniker and above all, Eric and Susie, Lord and Lady de Saumarez. My thanks also to David Lambert, John Popham, Christopher Ridgeway and Steve Thomas for kindly providing me with invaluable information. I have drawn extensively on the excellent work of Pat Murrell (on the Cullums) and Jon Phibbs (on Ickworth). The drawings were provided by Phillip Judge and the aerial photographs by Derek Edwards. Photographs of documents in the Suffolk Record Office were taken by Doug Atfield and paid for by Suffolk County Council. The cover photograph and Plate 14 are by Alastair Tuffill; the others are by myself or by Michael Brandon Jones of the University of East Anglia (UEA). I am especially grateful to the staff of the various branches of the Suffolk Record Office, who were always cheerful and helpful.

Lastly, I would like to thank the many past and present students at the Centre of East Anglian Studies at UEA, who have made aspects of Suffolk garden history the subjects of PhD theses, MA dissertations or undergraduate dissertations. I have drawn, unashamedly but I hope not without acknowledgement, on their work: my thanks go to Kate Bond, David Brown, Rosemary Hoppitt, Elise Perciful, Min Williams and Sally Wilkinson. Members of the Centre of East Anglian Studies provided, as ever, the best possible working environment: thanks must go, in particular, to Susanna Wade Martins, Edward Bujac, Hassell Smith, Richard Wilson, Mavis Wesley and Jenni Tanimoto. The publication of this work would not have been possible without a generous grant from the Ann Ashard Webb bequest.

Abbreviations

ESRO	East Suffolk Record Office, Ipswich
WSRO	West Suffolk Record Office, Bury St Edmunds
LRO	Lowestoft Record Office
PRO	Public Record Office, Kew

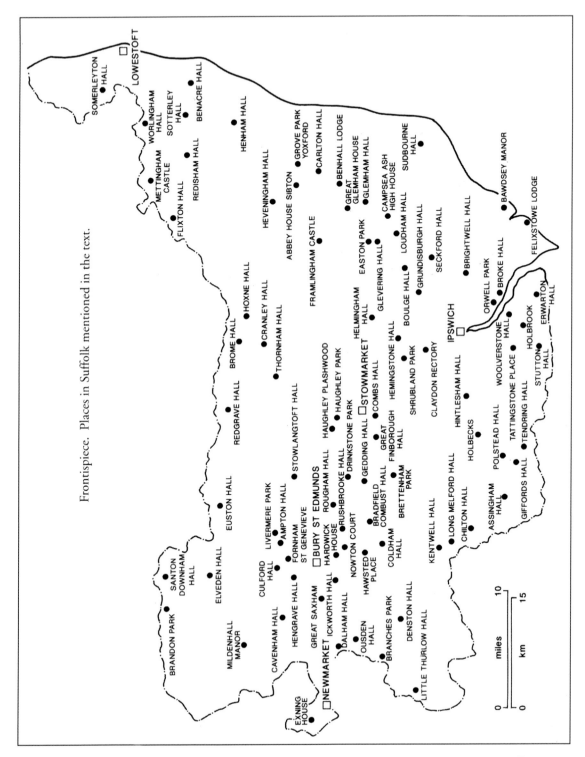

Frontispiece. Places in Suffolk mentioned in the text.

CHAPTER I

The context of garden design

This book is about the gardens, pleasure grounds and parks which were laid out around the houses of the gentry and nobility in Suffolk in the period between the early sixteenth and the late nineteenth century. It is not a comprehensive account of the county's garden history, and it deals only in passing with the gardens of other social groups, but it does examine the kinds of designed landscape large and durable enough to have survived to the present, and which thus still make an impact on the modern landscape. It is a work of landscape history rather than horticulture, and readers will, I hope, forgive my sometimes cavalier approach to matters botanical.

Students of garden history often discuss the development of garden and landscape design as if it was something that occurred in a vacuum. In this volume, parks and gardens are considered as part of the real world. Money had to be found for their creation and maintenance, and this makes it hard to discuss them without some examination of economic history. Their layout and design expressed the beliefs and attitudes of their owners and creators, so some attention must be paid to wider intellectual, philosophical, social and ideological developments. We need to note, in passing, changes in the style of Suffolk houses, for many and complex were the links between garden styles and architecture. Above all we must consider the wider landscape – the natural topography and the 'vernacular' countryside of fields, woods, farms and villages. For, as we shall see, this had a determining influence, especially on the character of the more extensive designs.

The landscapes of Suffolk

We must begin with soils and geology – not only because gardens contain plants, and plants need soil, but also because these things helped determine so much else that materially affected the development of designed landscapes in Suffolk. Underlying almost the whole of the county is a thick deposit of chalk, but this, dipping towards the south and east, is buried ever deeper beneath more recent formations. In the east and north-east it is obscured by 'Crag' deposits – a varied collection of Pliocene and Pleistocene gravels, clays and shelly sands. In the extreme south of Suffolk – straddling the border with Essex – other Tertiary formations occur: sands and gravels and London clay.

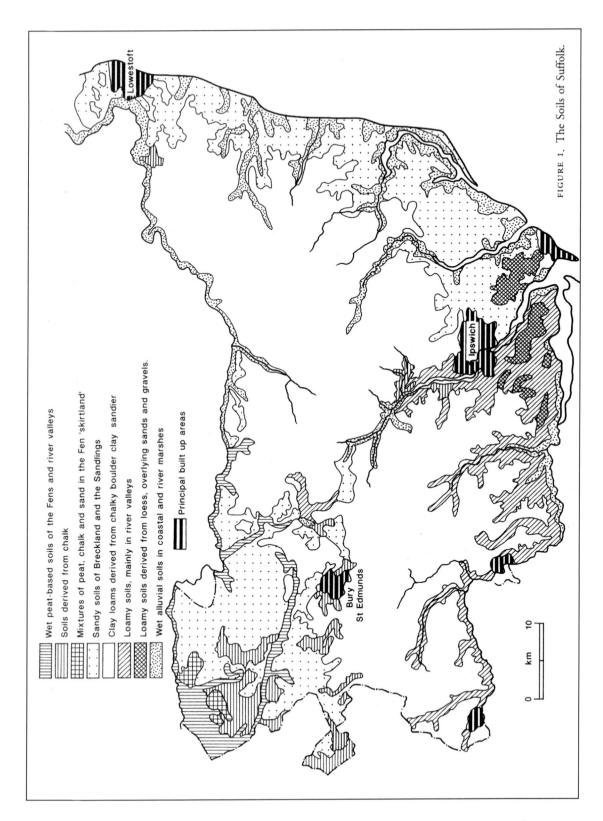

Wet peat-based soils of the Fens and river valleys

Soils derived from chalk

Mixtures of peat, chalk and sand in the Fen 'skirtland'

Sandy soils of Breckland and the Sandlings

Clay loams derived from chalky boulder clay sandier

Loamy soils, mainly in river valleys

Loamy soils derived from loess, overlying sands and gravels.

Wet alluvial soils in coastal and river marshes

Principal built up areas

Lowestoft

Ipswich

Bury St Edmunds

0   km   10

FIGURE 1. The Soils of Suffolk.

2

All this, however, is largely blanketed beneath a diverse range of glacial deposits. Across the centre of the county, in a wide band which extends south-westwards into Essex and northwards into Norfolk, chalk and crag are buried beneath a mantle of boulder clay, in places more than 250 metres thick. This forms a slightly tilted plateau, dissected to varying degrees by river valleys. The clays themselves are of varied composition, in places containing large quantities of chalk, but elsewhere, and especially towards the north-east, more sandy and acid in nature. To the north and west of this belt of clays, however, and to the south and east, the glaciations had different effects. Here glacial meltwaters, and high winds blowing close to the ice fronts, deposited a diverse range of sands and gravels which gave rise to an equally diverse range of soils, although these are, for the most part, acid and free-draining.

Breckland

'Breckland' is the term usually given to the area of light soils lying to the north-west of the clay belt. This is itself a modern term, first coined as late as 1925 by W. G. Clarke,[1] but the region has long had a distinctive identity, born of its agriculturally marginal character. Problems of low rainfall and sharp late frosts are compounded by the acid and freely-draining nature of the local soils. In the last decades of the eighteenth century, to judge from Hodskinson's county map of 1783, well over 40 per cent of the district was still occupied by heaths and warrens. The better soils, mainly located within the principal valleys, were the main arable areas, traditionally farmed in 'open fields' containing the intermixed, unhedged strips of many farmers. This ploughland was kept in heart by the dung of sheep flocks which, grazed by day on the open heath, were folded at night on the arable strips when they lay fallow or after the harvest.[2] The sheep were primarily valued for their role as mobile muck-spreaders, but landowners also profited from them more directly: in 1624 Robert Lord derived the staggering sum of £522 from his Elveden foldcourse (£31 for skins and pelts, £259 for hoggets and lambs, £83 for crones and £149 for 79 stone of wool).[3] At the start of our period Breckland was a largely hedgeless and treeless landscape, and a relatively underpopulated one. Indeed, some settlements on the very worst soils had dwindled in size, or even became deserted completely, in the late middle ages. Some became warrens or sheep-walks, with little or no arable – places like Wordwell which consisted of a single farm in 1736, 'the chief profits whereof arise from a flock of sheep, the soil being for the most part a barren dry heath – a very bleak place'. On the edge of the region, rather more extensive areas of the better, more calcareous soils could be found; here the area of heaths was less and the area of arable greater. But all of this region was largely open in the seventeenth century, hence its traditional name of 'Fielding'.

In the period after 1750, and especially in the decades either side of 1800 – when, because of the Napoleonic blockade, prices were abnormally high – landowners embarked on ambitious schemes of reclamation and improvement.

FIGURE 2. The Breckland landscape: one of the few remaining areas of open heathland, Knettishall.

Open fields and commons were enclosed, often by Parliamentary Act, and large areas of heathland were marled and ploughed. But local landowners were motivated as much by a fashionable enthusiasm for improvement as by a careful consideration of economic and agricultural realities. To some at least, reclamation was a civic duty, and part of a wider transformation of the Breckland landscape which included widespread tree-planting and the laying out of elaborate and extensive parks and pleasure grounds. In the cooler economic climate of the 1820s and 1830s, much of the reclaimed land went out of cultivation.[4]

The 'Woodland'

The clay belt which runs through the centre of the county does not constitute a homogenous zone. Although often grouped together as 'Woodland' Suffolk, there are important differences in soils, topography, and historical development between areas in the south and west, and the north and east, of this broad strip of countryside. Robert Reyce in the seventeenth century was well aware of this difference, distinguishing between the latter district – 'which cheifly

4

consist upon pasture and feeding' – and the south-west, the 'midle parts', which although 'enjoying much medow and pasture, yett far more tillage doe from thence raise their cheifest maintenance'.⁵

The north-east, where the clay soils were heaviest and the plateau most level, made good pasture land and was traditionally cattle-farming country. Open fields had existed – especially in the principal valleys – in the early Middle Ages but by the time our story begins this was an enclosed landscape, largely under grass, with many hedges, woods and copses. Settlement was often highly dispersed, with numerous isolated farms and with hamlets clustering beside greens and commons. Trees were numerous, both in hedgerows and in 'rows' – thin strips along the field margins. At Denham, a farm survey of 1651 suggests that there was an average of 15.4 trees per acre, while another, made for an estate in Thorndon in 1742, implies a density of no less than 29 per acre: 67 per cent were oak, 16 per cent ash and 17 per cent elm. In all, 82 per cent were pollards, only 13 per cent were classified as saplings and a mere 5 per cent were timber trees.⁶

This 'traditional', bosky landscape changed in a number of ways in the course of the eighteenth and nineteenth centuries. From the 1760s there was a gradual expansion of the area under the plough, and a gradual reduction in the importance of bullocks and dairies, a trend which intensified in the Napoleonic War years. By 1840, when the Tithe Award maps were surveyed, the claylands was a largely arable region, with significant areas of permanent pasture only remaining where soils were particularly tenacious. Whereas in 1650 most farms in the area had, on average, perhaps 20 or 25 per cent of their land under the plough, by 1840, 70 or even 75 per cent was the norm. These changes were associated with some reduction in the numbers of trees and hedges, but this nevertheless remained – until the devastation wrought by modern agricultural intensification – a comparatively well-treed and well-hedged landscape.

The south-western claylands have a rather different topography and a rather different history. Here the landscape is more rolling, the clay soils more calcareous and better drained, and consequently – as Reyce suggests – more land was under arable cultivation at the start of our period. This, too, was an enclosed landscape by late medieval times, and more densely wooded than the north-east, although there were probably fewer farmland trees.

The Sandlings

To the east of the clay belt lay another district of poor, acid soils, the 'Sandlings' or 'Sandlands'. Here, as in Breckland, the thin arable soils were kept in heart through the systematic night folding of sheep grazed by day on extensive tracts of heathland. Here, however, the extent of poor soils was less: they formed a comparatively thin strip between the clay uplands and the sea. Many parishes also possessed areas of coastal marsh, which were progressively embanked and improved in the course of the post-medieval period. Others extended up onto the claylands to the west, and thus included areas of fertile if heavy soils. Some parishes, especially towards the north, contained portions of all three environmental zones – clay, sand and marsh.[7]

In the Middle Ages the Sandlings had been a largely open landscape, but enclosure proceeded faster here than in Breckland, and by the start of the eighteenth century open fields were already in retreat and much of the heathland was privately owned. Here, too, the second half of the eighteenth century, and in particular the Napoleonic War years, saw much reclamation and improvement. In 1795 Young described how he had recently crossed the 'extensive wastes of Sutton', commenting:

> Having long ago called on the farmers publicly to cultivate them, I cannot but recollect the answers I then received – that it would not answer – and that they were fit only for what they gave – coarse sheepwalk. I have now the pleasure to find my old opinion confirmed, for great tracts have been broken up within these twenty years, and are found to answer well ...[8]

Nevertheless, conversion was by no means total, and many of the greatest heaths, such as Foxhall, and Martlesham, survived largely untouched into the twentieth century. Many smaller areas of heath continued to exist within the bounds of ring-fence farms.

The genius of place

The relevance of this brief foray into Suffolk's landscape history will hopefully become apparent in the course of this volume. It is sufficient here to say that variations in soils had some effect on the kinds of trees and other plants which could be successfully grown, while variations in the character of the working countryside – and especially in the numbers and antiquity of hedges and trees – determined the nature of the raw materials with which the designers of parks, in particular, had to work. But what eighteenth-century writers referred to as the 'genius of the place' was also a matter of topography – of the configuration of landforms – and this, too, was in part related to geology. Suffolk can hardly be called a hilly country, but it is for the most part gently rolling, and only in two districts can extensive areas of completely flat terrain be found: in the Fens of the north-west, an area shunned by the fashionable in all periods; and in the north-east, on the level clay plateau of 'High' Suffolk, especially in the area between Halesworth and Bungay. Nevertheless, the generally muted nature of the terrain often posed problems for designers. Lakes were hard to construct where, as was often the case, valleys were wide and their gradients gentle; only in Breckland – paradoxically, the driest part of the county – are large bodies of water a fairly common feature, in parks like Culford, Livermere, and Ampton. Cascades and fountains were always hard to construct – Suffolk has nothing to compare with the seventeenth-century waterworks at Broughton or Chatsworth – while the absence of high hills precluded the creation of rugged 'picturesque' prospects, and affected the nature of views and vistas more generally. As the great designer Humphry Repton succinctly put it, 'The general flatness of the County of Suffolk will not allow of romantic scenery, or very extensive prospects.'[9] There are, however, some exceptions to this general rule, most notably at Shrubland Hall to the north of Ipswich, where, as we shall see, a series of designers made great use of a steep escarpment on the edge of the Gipping valley to provide an impressive setting for the hall.

Landowners and estates

The effects of variations in soil and topography on the design and character of Suffolk's parks and gardens is complicated by the fact that the homes of the wealthier members of society were not equally distributed across the county. As Richard Blome observed in 1673, 'High Suffolk or the Woodland is chiefly the seat of the Yeomanry, few being there either very rich or very poor ... In the great Towns, the mixt soil, the fielding of Bury and the Sandlands, the Gentry are commonly seated.'[10] By 'mixt soil' Blome seems to have meant the lighter, more dissected clays, in the south-west of the county, and to some extent on the edge of the clay belt elsewhere. This pattern is very evident in the distribution of large houses – those with ten or more hearths – listed in the Hearth Tax of 1674.[11] Where large houses did exist within the main clay belt, they were usually associated with ribbons of light soils in the principal valleys, and especially those of the Waveney and the Gipping. Obvious examples include Flixton, above the Waveney – owned during the eighteenth century by the Wyburns and from the 1750s until the late nineteenth century by the Adairs; and the Shrubland estate, in the Gipping valley, owned until 1770 by the Bacons and subsequently by the Middletons. The extent to which small landowners flourished on the clays should not, however, be exaggerated. Much of the land here had, by the early eighteenth century, been acquired by large estates, but these were normally based elsewhere, and their properties were often splintered, rather than forming large consolidated blocks.

The explanation for this broad relationship between soils and land-ownership patterns is not entirely straightforward. In part it was the consequence of very ancient factors – the heavier clay soils had, even in the Middle Ages, boasted substantial numbers of free men, largely independent of manorial control. In part it was the result of farming patterns which developed – or at least, became more sharply defined – from the fifteenth and sixteenth centuries, with the lighter lands specialising in grain production, the heavier soils in cattle rearing and dairying. Over time, in all areas of England, arable pursuits tended to favour the emergence of a social pattern based on large estates, large tenant farms and landless labourers, while pastoral pursuits, in contrast – and dairying especially – generally produced a less hierarchical society with a plethora of small farms and small, 'gentry' estates.[12] The main reason why large estates flourished on the lighter lands is, however, probably rather simple. In all periods, the poor soils of the Sandlings and Breckland commanded a much lower price than the heavier, but more fertile, clays. As a result, it was much easier to build up a large, compact holding in these regions than it was on the claylands of the Woodland.

Whatever their cause, variations in patterns of ownership had an obvious effect on the distribution of large designed landscapes. They tended to be found where the rich resided and where large continuous estates existed, and not where properties were smaller or more splintered. As Repton – ever an acute observer – noted in his *Red Book for Culford Hall,* in many parts of Suffolk 'the Character of great dignity or magnificence is seldom attainable, because it requires undiverted command of territory incompatible with the kind of Neighbourhoods in which the county abounds.'[13]

Location and geography

Suffolk lies on the eastern side of England and like other counties so located has a markedly continental climate, with dry summers, severe winters and sharp late frosts. Droughts could destroy young trees and plantations, and tree-planting was a difficult task – although one pursued with dedication – on the dry soils of Breckland or the Sandlings. Winter weather could likewise cause tremendous damage, especially to the less hardy plants in garden and pleasure ground. In 1795 the Rev. Sir Thomas Gery Cullum of Hardwick mused on the freezing point of sap in trees, noting that:

> We cannot be surprised that some of our severe seasons such as this should destroy some of our most beautiful winter-flowering shrubs such as the Abutus, and Laursetinus, when the Thermometer has several times been lower than 15, and more than once been down to 5 and 4.[14]

Many exotic trees which thrive in the western half of England do badly in the east, and many of the species planted in arboreta and pineta in the nineteenth century have failed to survive to the present.

Suffolk's location within England had a number of other effects, often subtle, on the development of parks and gardens. It was sufficiently remote from London in the early eighteenth century to have been in some ways rather marginal to the key developments in garden design. By the end of that century, however, improvements in transport ensured closer economic contacts and some influx of London money. In addition, an absence of good supplies of water power or of the other natural resources required by the industrial revolution ensured that the latter phenomenon largely passed the county by. It remained essentially rural, even in the nineteenth century, as the indigenous textile trade suffered a gradual but inexorable decline. For much of the period studied here, the county had, in effect, a de-industrialising economy, and labour costs were as a result comparatively low – especially in the decades either side of 1800. Gentle terrain and fertile soils ensured a prosperous agriculture, especially in the boom years of the late eighteenth and early nineteenth centuries, when much of the former grass land on the clays, and the great heaths on the light land, went under the plough

FIGURE 3. Suffolk
under the plough,
1816. This painting by
John Constable – a
preliminary sketch for
his 'Mill on a
Common' – shows
East Bergholt
Common being
ploughed, probably
for the first time,
following enclosure.

(Figure 3). What Repton wrote of the area around Glemham in 1791 was true
of many parts of Suffolk:

> The natural Tameness of the country is amply compensated by a dry soil
> without sterility, and an easy communication from excellent roads; to
> which may possibly be attributed the general air of cheerfulness which
> prevails throughout the whole district ... the country appears everywhere
> fully inhabited, yet not crowded ...[15]

FIGURE 4. Thornham
Hall: the entrance
front, as depicted on
an early seventeeth-
century painting.

10

CHAPTER 2

Parks and gardens before *c.* 1660

The earliest gardens

Since the early Middle Ages the homes of wealthy landowners in Suffolk had had ornamental gardens and pleasure grounds laid out beside them, and the gardens of Tudor times had a long pedigree. By the start of the sixteenth century most comprised a number of distinct areas enclosed by hedges of hawthorn or box. Arbours (seats or walks covered by trellis-work or hedging), seats of turf and herb gardens were all popular features. The College beside Mettingham Castle was surveyed in 1562, some time after its Dissolution by Henry VIII. Its garden was

> Sett with diverse trees of fruite and devided into sondrye partes with quicksett hedges and quicke hedges of boxe where hath byn manye fayer Arbors and many small gardens ... and hathe fower little pondes in it called fridaye pondes [which] served to preserve fishe taken on weken dayes tyll fridaye.[1]

In the course of the sixteenth century 'knots' (geometric, interlaced patterns, defined by low hedges of box, thyme or hissop) became fashionable in the gardens of the gentry. Sir Thomas Kytson of Hengrave brought a 'Dutch-man' from Norwich to 'viewe ye orchardes, gardyns and walkes', and later paid him for 'clypping the knotts, altering the alleys, setting the grounde, finding herbes, and bordering the same', in the 1570s.[2] The hall is shown on a map of 1587 but this is unfortunately too schematic to provide much information about the layout of the gardens.[3] Some early maps, however, are more informative. What is apparently a knot or simple *parterre* appears as a tiny detail on a map of Brundish manor, surveyed in 1627, and some kind of geometric design is shown on a map of Hoxne Hall, made in 1619, within an area labelled 'the garden', although its character is obscure (Plate 3). Similar areas appear to the west of the house shown on a map of 1626 (surviving only as a later copy) of Cranley Hall in Eye.[4] These maps – and others, such as that of Badingham Hall, 1614 – suggest that the main areas of ornamental garden usually lay beside the parlour – the main living room of the owner.[5] On the entrance fronts, other kinds of arrangement could be found. An undated painting of *c.* 1620 of the entrance front of Thornham Hall (Figure 4) shows two plain lawns, separated by a path leading to the front door, each with a central flower bed, specimen conifer and flower borders.

The surrounding walls are surmounted by flowers in pots, placed at regular intervals.[6]

Ornamental gardens generally occupied, as in later times, one of several enclosures and facilities clustered around the hall. Sir John Hopton's Cockfield Hall had, in the late fifteenth century, a bakehouse, barn, and dovecote in its immediate vicinity, while beyond these lay the fishpond, garden, and park.[7] The various enclosures or yards were bounded by fences, hedges or – increasingly as the sixteenth century progressed – brick walls. Where the soils were heavy and water retentive – that is, on the clay belt running through the middle of the county – or where houses were located in low-lying positions close to rivers or streams, all or part of this residential complex might be surrounded by a wide, water-filled ditch or moat. As Reyce put it in 1618, 'all our antientest houses, for their more security and quiet against all worldly accidents, were ever so placed that they were environed with a broad ditch or moat'.[8] In reality, moats were as much about ornament as defence: few would have deterred anything more than a casual intruder.[9] They conferred a range of practical advantages, including enhanced drainage, but in large measure they were a medieval fashion statement, proclaiming the martial aspirations and therefore superior status of the house's owner. The medieval mind clearly found something particularly appealing about a mansion surrounded by, and reflected in, an area of water, and well into the period covered by this book moats continued to be maintained with enthusiasm.

In this sense moats were, at least in part, garden features in their own right. But they also often embraced the ornamental gardens beside the hall. Sometimes, as at Hoxne or Barrow, a single moat surrounded both residence and garden.[10] There are numerous other examples on later maps, almost certainly showing arrangements unchanged in their essentials since the sixteenth century. A finely-detailed map of Bedingfield Hall, for example, made in 1729, shows the moat surrounding an area of four acres (*c.* 1.6 hectares), containing not only the house but also garden, outbuildings, fish ponds and orchards.[11] A number of elite residences erected in the county in the fifteenth or early sixteenth centuries, however, appear to have been provided with double moats, one occupied by the house and yards, the other perhaps by gardens. At Helmingham, John Tollemache – whose family had acquired the estate in 1487 – began to build a fine new hall on a moated site in 1509. The adjoining moat almost certainly surrounded, from the start, an area of garden, although the walls which now stand here appear to be of mid-eighteenth-century date, and the details of the planting are modern.[12]

Water thus played a key role in the gardens of the fifteenth and sixteenth centuries, and earlier too; indeed, it has been suggested that the large areas of water found around or beside such medieval castles as Bodiam in Sussex or Kenilworth in Warwickshire were, at least in part, similarly aesthetic in character – part of large-scale, elaborately designed landscapes intended to set off the castle to good advantage.[13] At Framlingham Castle in Suffolk, the location of the two large lakes or 'meres' is certainly suggestive.

Parks and gardens before c.1660

They lie beside the castle, on the same side as the principal hall range. Immediately below the latter is an earthwork enclosure which may have originated as a bailey of the original castle on the site, but which was occupied by a garden in the sixteenth century and probably also in medieval times (there are references to gardens at the castle in 1302, although their precise location is unclear).[14] The mere lay beyond this, and would thus have formed a striking backdrop to the view from the principal rooms of the residential block. Recent archaeological investigations have demonstrated that the meres are, at least in part, of natural origin,[15] but this need not preclude their use, and perhaps adaptation, as an aesthetic landscapes feature during the later Middle Ages.

In the course of the sixteenth century, as bricks came into more common use, walls were increasingly used to enclose gardens. They were often embellished with crenellations and provided with impressive gates. Walled gardens fell out of favour, as we shall see, in the course of the eighteenth century, and most of these enclosures were demolished, but a number survive in Suffolk, although comparatively few from the period before 1660. One of the finest examples is at Stutton Hall, overlooking the Stour estuary in the far south of the county. The hall was built by Sir Edmund Jermy in 1553, and the gardens were presumably laid out at the same time – the brickwork of the house and garden walls is virtually indistinguishable.[16] Typically, their survival here is due to the fact that by the eighteenth century the hall had declined to the status of a substantial farm house. The walled garden adjoins the hall to the north, and nineteenth-century maps suggest that it was originally mirrored by another lying on the far, southern side of the hall, probably removed in the 1890s when the house was refashioned by James Fison. Its walls are built of thin but regular bricks laid in English bond, with piers spaced at intervals of *c.* 12 metres, surmounted by elaborate finials similar in design to the chimneys of the house (although some, if not most, appear to be later replacements). A gate is set in the centre of the north wall. It is an elaborate feature, again ornamented with finials (Figure 5). The outer entrance arch – on the north side – has a low, 'perpendicular' arch of normal Tudor form. The inside doorway, however – i.e. that on the southern side of the wall, facing the house – has a round classical arch with flanking fluted pilasters, a clear statement of Jermy's Renaissance pretensions. The northernmost 7 metres of the eastern and western walls are raised by *c.* 1.5 metres, as is the whole of the north wall. Inside the garden the ground level increases correspondingly towards the north. There must once have been a low terrace on this side of the garden, allowing elevated views across the knots and other geometric planting inside.

Equally impressive remains survive at Melford Hall at Long Melford, built in the mid-sixteenth century for Sir William Cordell. This substantial, brick-built house, with turrets and a fashionable long gallery at first-floor level, is ranged around three sides of a square, although a plan by John Thorpe (Soane Museum) suggests that there was an intention at one point to close

FIGURE 5. The
sixteenth-century
gateway at Stutton
Hall.

off the fourth side.[17] The main front lies to the east so that the house
turns its back on the adjacent village green. Probably contemporary with the
hall is the fine octagonal summer house, with its eight prominent gables and
stumpy pinnacles, which stands at the western end of a substantial terrace
running east–west to the north of the house (Figure 6). This perhaps functioned
as a 'banqueting house' a common feature of sixteenth- and seventeenth-
century gardens – a small building in which a banquet was eaten. This was
not the large, sumptuous feast which the modern use of the word might
imply, but an elaborate final course to a meal which had been consumed
elsewhere, in the Great Chamber or, perhaps, in the hall of the mansion.
Fine maps of 1613 and 1615[18] show that the hall lay within elaborate formal
grounds enclosed both by walls and moats. It is noteworthy that although
the mansion turned its back on the adjacent village, the banqueting house
was positioned in the wall that separates the gardens from the public road.
Such a position would have allowed the Cordells and their guests to sit and
enjoy, at one remove, the comings and goings around the village green.
Interestingly, the fine sixteenth-century summerhouse/banqueting house at

FIGURE 6. The
octagonal summer
house, Melford Hall,
Long Melford.

Seckford Hall near Woodbridge, a tall building with stylish stepped gables, is similarly placed.

Moats gradually fell out of favour as a setting for a residence in the second half of the sixteenth century. When Sir Nicholas Bacon built his new house at Redgrave in the 1540s, there was no moat and the gardens consisted of two enclosures surrounded by walls, again embellished with turrets or pinnacles. The enclosure to the rear of the house measured around 50 metres square and was divided into two parts: the half nearest the house comprised an orchard of fruit trees dissected by walks or *allées*, while the other had a central pond surrounded by a brick terrace. The other enclosed garden lay to the side of the house and was bounded on one side by a detached loggia with a timber gallery above.[19] Nevertheless, existing moats were retained, and owners evidently continued to find them appealing. At Chilton near Hadleigh the early-sixteenth-century hall (much reduced from its original size) stands within a substantial moat but the walled garden, its brickwork indicating an early seventeenth-century date, lies outside and to the west of this.[20] There does not ever appear to have been a wall on the eastern side of the garden adjacent to the moat. Traces of a terrace walk here and the (much altered) remains of contemporary covered seats built into the eastern ends of the northern and southern walls suggest that views across the moat were still relished by the hall's owners, the Crane family.

Walled garden enclosures of sixteenth- and early seventeenth-century date survive at a number of other places in the county, although usually in fragmentary form – as at Hengrave, where the much-altered remains are incorporated into walled gardens of mainly eighteenth- and nineteenth-century date; or at Little Glemham, where the western and parts of the northern wall are of sixteenth-century date, while the others appear to have been built around 1720, when a systematic rebuilding of the hall was accompanied by a comprehensive remodelling of its grounds (see pp. 41–2).

During the late sixteenth and seventeenth centuries a number of English gentlemen and aristocrats created gardens modelled on those of contemporary Italian villas. Some of these individuals had actually travelled to Italy and experienced these landscapes at first hand, for the Grand Tour was not entirely an invention of the eighteenth century. Italian Renaissance writers like Alberti argued that gardens should be planned as an integral part of the architecture of the house; in particular, they should be arranged symmetrically in relation to one of its main façades. These gardens made considerable use of raised terraces, not (as probably at Stutton) to enhance views within an enclosed garden area, but to provide a panorama across the wider estate land. Their designers also put considerable emphasis on variety: Italian writers urged that there should, ideally, be a number of different garden areas, allowing scope for exploration and opportunities to surprise the visitor. These were organised in such a way that those nearest the house were of a structured, geometric character, while those further away became increasingly irregular, with the most distant areas perhaps taking the form of a 'grove' or 'wilderness' – a woodland garden. The need for overall symmetry ensured that the principal garden areas were often arranged on either side of a single axial walkway leading out from the mansion. Lastly, Italian gardens usually made use of statues and other cultural debris looted from local archaeological sites, and often included water features – ponds, cascades and so on – and grottos or similar cave-like features.[21]

Somerleyton Hall in north-east Suffolk is today justly famous for its nineteenth-century gardens, but its seventeenth-century grounds must, by the standards of the time, have been even more impressive. We are fortunate in having a fairly detailed plan of these remarkable gardens, included on an estate map of 1652 (Figure 7), and a written description made to accompany this in 1663 by the surveyor Thomas Martin.[22] The map clearly marks the location and layout of 'the manor of Somerleyton the gardens, orchards, and Firrendale yard, woods and walks'. The hall had a main block facing west and two wings ranged east–west: this is the house that survived, with some alterations, until rebuilt by Morton Peto in the 1840s. An orchard lay to the south and two enclosed courts to the west. The main gardens, however, lay to the north of

FIGURE 7. The gardens and park at Somerleyton Hall, 1652.
COURTESY SUFFOLK RECORD OFFICE

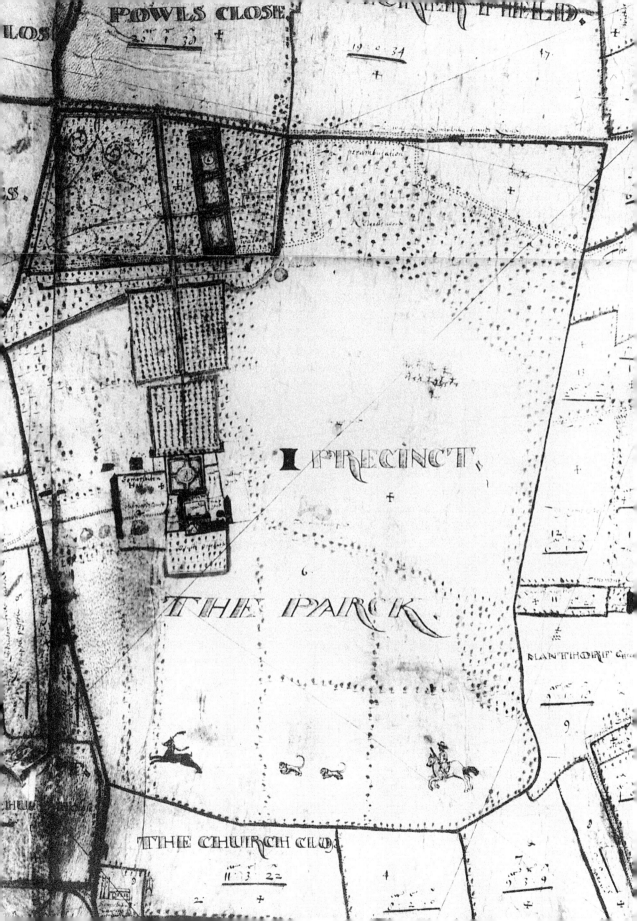

POWLS CLOSE

I PRECINCT

THE PARCK

THE CHURCH CLO.

the house. There were four distinct areas, strung in a line along a central north–south walk, which increased in size with distance from the hall. First in the sequence came the Great Garden, its geometric layout typical of the period. It was square, with four paths running in from the sides, meeting at a central circular feature. On its northern side it was bounded by a raised 'terrace walk', with banqueting houses at each end. North of this lay the North Orchard, presumably an ornamental area as much as one devoted to the production of fruit. This in turn was bounded by the third main garden enclosure, the Firrendale Yard, a more unusual feature, described in the 1663 survey as:

> Being of late the most incomparable piece in the Realm of England but now ruinated by a great wind it being planted in the year of Our Lord God 1612 with 256 fir trees, most part of them before they were blown down were above twenty yards in height with a very fine walk in the midst of them.

Thomas Fuller, writing in his *Worthies of England* of 1662, was particularly impressed with this area, describing:

> Somerleyton Hall ... belonging to the Lady Wentworth, well answering the name thereof: for here *summer* is to be seen in the depth of *winter* in the pleasant walks, beset on both sides with fir-trees green all the year long, besides other curiosities ...[23]

To the north of the Firrendale Yard came the final, and in many ways the most remarkable, area of the grounds. The 1663 survey describes it as 'The Woods and Walks with variety of seats, statues, fish ponds, a house for pleasure newly erected and diverse other rarities'. The map shows that this area, unlike the others, was not laid out in a rigidly geometric fashion. Instead, it included serpentine walks threading through the trees, while the main features – the line of three ponds – lay to one side of the main north–south walk, in an asymmetrical fashion (not mirrored by similar features on the other side). The map also shows that one of the 'diverse other rarities' in this area was a grotto to the north of the ponds.

While the gardens were still under construction in 1619 William Edge, the bricklayer and probable co-architect of Raynham Hall in Norfolk, sent a message to his master Sir Roger Townshend (via Thomas Barker, Townshend's stewart) describing how he had visited Somerleyton:

> And sawe the works the carver ther had practised and done in earth, both burnt and not burnt, the statures both of men and women, setting uppon several beasts, and are to stand uppon great pillers made of brickearth and fytting to stand att the meeting of a crosse walke in a woode or otherwise. And there is more the xii signes and xii other strange beasts which are to be seated in the gardinge uppon severall pillars.[24]

He also mentions the waterworks he had seen there, no obvious sign of which appears on the map – a good indication of the limitations of our most important source for early garden history. Quite how and why the Wentworths – not a family of particular distinction, although (as the 1653 map also shows) masters of a sizeable estate in Lothingland – came to lay out this precocious garden remains unclear. There were doubtless others, of similar sophistication, in early Stuart Suffolk, but as yet we know nothing of them.

Parks

The Somerleyton map shows that the hall and gardens were set within a park. 'Park' is a term which requires some explanation, for it has changed its meaning gradually but significantly over the long centuries. In early medieval times parks were primarily places where deer were kept; they were hunting grounds and also, more importantly, venison farms. They were also used to produce timber, and sometimes for grazing other forms of livestock. They were, in essence, residual fragments of the once extensive 'wastes', now enclosed and preserved by manorial lords. They were usually quite densely wooded. Some contained areas of embanked coppiced woodland, but they were principally 'wood-pastures', that is, areas containing timber and pollarded trees which were spaced widely enough to allow some grass to grow between and beneath them. They normally had only limited areas of more open ground, referred to as 'launds' (the main denizens of medieval parks, fallow deer, flourish best in a well-timbered environment). The only place in Suffolk where it is still posisble to savour something of the flavour of a medieval deer park is at Staverton Park, near Wantisden, north of Woodbridge. Here an area of over eighty hectares is still covered with closely spaced, ancient pollarded oaks – a unique survival, and a perplexing one, as the available evidence suggests that the park ceased to be stocked with deer more than four centuries ago.

The most important elite residences in medieval times – especially great castles like Framlingham – often had deer parks immediately adjacent, which may already have been landscaped to some extent in an aesthetic manner. But most medieval parks were found at some distance from manorial residences, often on or close to parish boundaries, and many therefore contained a 'lodge' which served as a residence for the park-keeper and sometimes provided temporary accommodation for the owner when hunting took place in the park.[25]

Suffolk's medieval parks have been studied in considerable detail by Rosemary Hoppitt (Figure 8).[26] Her research has demonstrated that their numbers increased gradually in the period up to 1250, and then very rapidly in the second half of the thirteenth century, a time of considerable economic expansion. By 1350 there were probably more than fifty parks in the county; most were located on the belt of heavy clay soils running through the middle of Suffolk, largely because it was here that significant quantities of woodland survived into the high Middle Ages. The numbers of parks dwindled in the wake of the Black Death, however, and by the second half of the fifteenth

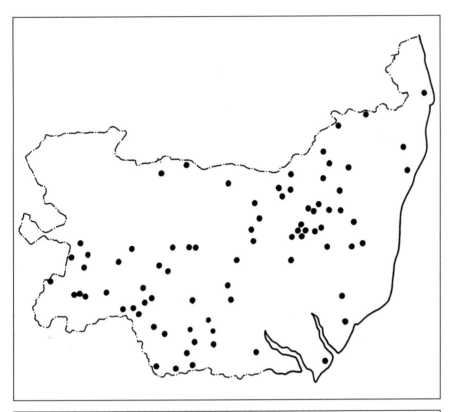

FIGURE 8. The
distribution of parks
in Suffolk, 1066–
*c.*1400 (after Hoppitt
1992).

FIGURE 9. The
distribution of parks
in Suffolk, *c.*1400 to
*c.*1600 (after Hoppitt
1992).

century there were probably only around thirty-five still functioning in the county. This reduction in numbers was partly the result of the decline in population, which led to escalating wage costs which made the upkeep of the park, and especially of the perimeter 'pale', or fencing, ruinously expensive. But it may partly have been, as Hoppitt has suggested, because deer herds were decimated by the same diseases that destroyed many cattle herds at this time.[27] Either way, in the fifteenth century the number of parks began to increase once again, and by the early seventeenth century it had more than returned to their early medieval peak.[28]

What is particularly interesting, however, is that in the period after 1350 the location of parks gradually changed.[29] Most of the parks that disappeared in the late Middle Ages were in remote locations. Most of those that survived were located beside residences, as were almost all of the new parks that appeared in the course of the fifteenth and sixteenth centuries. These changes in the location of parks were accompanied by an alteration in their distribution within the county (Figure 9). As we have seen, by the seventeenth century most large estates were located on the lighter soils: either on the edge of the Sandlings, on the edge of the Fielding in west Suffolk or on the more broken, dissected claylands of the south-west. Parks gradually became more and more clustered in these areas. Early parks had mostly been established in areas of unenclosed manorial waste. These later creations – like almost all parks subsequently created within the county – were made at the expense of enclosed farmland. A number of the new parks of the fifteenth or sixteenth centuries were created in relatively poor and depopulated parishes, including Henham and Culford.[30] 'These were ideal places for rising and affluent families to acquire ... because land could be bought and re-organised, and the risks of opposition were not high.'[31] Emparking could lead to the further, and sometimes terminal, decline of settlements like these, and at Henham the church was left isolated within the park, as it was at Hengrave. Other parks, however, were associated with flourishing settlements. Several were established on the edge of expanding towns, by men grown wealthy in trade or the professions – like that at Woodbridge, created by the successful lawyer Thomas Seckford, or those laid out around Kentwell and Melford Halls on the edges of Long Melford.[32]

Those parks which fell out of use suffered a variety of fates. A few, like Staverton Park, continued to exist as recognisable features of the landscape, but were now managed as woods rather than as deer farms.[33] Most, however, were converted to farm land. The change in use and location was a gradual one, which continued into the seventeenth century. The three parks at Hundon, for example – Great Park, Broxted Park and Eastry – had become Crown possessions in the mid-fifteenth century, but were maintained as densely-wooded parks in the old manner and leased out to various families until 1611, when they were granted to William, Lord Cavendish, Earl of Devon.[34] To judge from a map made in the same year, they were immediately disparked, the trees felled, and divided into closes.[35]

As parks came to be created in close proximity to residences, their function and appearance began to change. They still contained deer and were still usually managed as venison farms and used for hunting. But they were now valued at least as much for their appearance.[36] This changed accordingly: early parks, as we have seen, had usually been fairly densely timbered. Later parks were more open – they began to develop that familiar appearance of turf and scattered trees which we today associate with the term 'parkland'. Few men wished to live within or beside a large wood. A more open although still sylvan landscape provided a suitable setting for the house and also allowed the size of the park – and thus the waste of resources employed in its creation and maintenance – to be proudly displayed.

Some parks had ornamental features constructed within them. Sir Nicholas Bacon seems to have built a viewing mount in the park at Redgrave in the 1560s.[37] The mound, now topped by an eighteenth-century obelisk, in the parkland 1.5 kilometres to the west of Helmingham Hall may also have originated as a viewing mount of sixteenth-century date, while the denuded remains of what may have been another feature exist at the end of the avenue at Hengrave, planted in *c*. 1587 (the 1839 Tithe Award map for Hengrave marks a feature called 'Mount Hill' at this point).[38]

A few examples may serve to demonstrate the changing character of parks in the late Middle Ages. The Bishop of Norwich had had a park at Hoxne in the claylands of north Suffolk since the thirteenth century. It was located in the area around what is now Park Farm, on the higher, heavier ground away from the village. In 1472, however, a lease of land refers to a parcel abutting on 'le Oldepark', implying that a 'New Park' had come into existence. This was close to the village, on lower and more undulating ground, and its creation may indicate the construction of a new bishop's palace here.[39] A map of 1619 shows the 'new' park and the manor house to one side of it – positioned, as was usual, within a moat and approached through a gatehouse. The map suggests that the park was largely composed of open pasture and scattered trees, with belts of trees around the perimeter.[40] No deer are shown, but the park was evidently fenced with a pale.

Long Melford was a similar case. A park is first mentioned here shortly before 1400,[41] but this was abandoned when Melford Hall was erected on the edge of the town in the mid-sixteenth century and a new park laid out around it. This is shown on a map of Melford manor, surveyed in 1580.[42] The park may have disappeared for a time in the late sixteenth century, however, for it is not mentioned in the Suffolk *Chorography* of 1602, and was newly licensed in 1613.[43] Sir Thomas Savage was granted a licence to impark '340 acres of park and warren around Melford Hall with the deer therein and full rights of chase and warren'.[44] The park extended away to the east of the house, and was clearly rather larger than the later landscape park here. The 1613 map shows the pale, and deer grazing within it (Plate 4).[45]

Similar in date is the park at Hawstead. In 1505 Sir Robert Drury acquired the manor of Hawstead and engrossed a number of other properties in the village. The development of the park was part of a wider scheme of aggrandise-ment which included alterations and improvements to the old Bokenham's Hall, the name of which was changed to the more grandiose Hawstead Place, and the abandonment of Hawstead Hall, which had been the family's main residence since 1464. The park extended across the northern part of the parish, occupying around 500 acres (200 hectares).[46]

The park at Hengrave, on the edge of the treeless Breckland, was a slightly later creation. Thomas Kitson, a wealthy cloth merchant whose family hailed from Lancashire, acquired the manor from the Crown in 1521 and began to build a new hall there in 1525. This was visited by Queen Elizabeth in 1578 and during the visit Thomas Kitson (son of the first Thomas) was knighted, and given licence to empark 300 acres (*c*.120 hectares) of land. A striking map of 1588 shows Hengrave Hall and its surroundings in the following year.[47] It shows the recently created park pale cutting through the surrounding agricul-tural land, which still largely consisted of hedgeless open fields. A substantial avenue is shown, running south-eastwards through the park from the southern façade of the hall. This is the earliest known avenue in Suffolk (and one of the earliest in England, although it is not entirely clear on the 1588 map whether the feature was created already, or being proposed).[48] The park was stocked with deer almost as soon as it was created; the Suffolk Record office holds a 'booke of accompt for all kyndes of Deere put to Hengrave which parke was fynished at Mygellmas in Anno 1587'.[49]

Parks very similar to these continued to be established in Suffolk into the seventeenth century. Helmingham Park, for example, almost certainly origin-ated in the first half of the seventeenth century: it does not appear in a list of Suffolk parks, compiled in *c.* 1560,[50] nor on Saxton's map of Suffolk made in 1575; nor is it mentioned in the description of the county, generally known as the *Suffolk Chorography*, compiled in *c.* 1600.[51] A rental of 1631 mentions an 'Old Park' of 122 acres, and a 'New Park' of 34 acres, but the latter must have been a long defunct deer enclosure which did not develop directly into the present Helmingham Park, for another rental (of 1651) contrasts the 'Old Park lately converted into a farm' with the 'new Park lately made about Helmingham Hall'. A rental for 1729 informs us that the New Park covered 120 acres (*c.* 50 hectares).[52] Somerleyton Park was also apparently a new creation of the early seventeenth century. It covered 130 acres (53 hectares) and was, according to the survey of 1663, 'very well replenished with deer and very full and excellent woods and timber lying in diverse parcels and angles'.[53] The latter slightly obscure phrase may relate to the fact that the trees within the park are shown growing in straight lines, clearly derived from hedges which had been removed when the park was laid out (Figure 7).

Not all sixteenth- and seventeenth-century deer parks in Suffolk consisted of wide expanses of open pasture. Some, even though located beside great houses, continued to have a more subdivided, less obviously aesthetic appear-ance; they were simply areas of woods and hedged fields enclosed by a stout, deer-proof fence. A map of Hawstead made in 1616 thus shows that the Great

Park was divided by fences and hedging into seven separate sections (two of them woods) and the Little Park into no less than ten.[54] As we shall see, some archaic, rather functional parks of this type continued to exist into the eighteenth century, but they were a minority. By the middle of the seventeenth century most parks were like Somerleyton or Hoxne – areas of open pasture with comparatively few subdivisions, scattered with trees and with fairly limited areas of woodland.

Although parks were now much valued for their appearance and increasingly ornamental in character, they nevertheless continued to have economic roles, producing timber as well as venison. Many also contained fish ponds, an indispensable adjunct to every gentleman's house even after the religious changes of the Reformation made the consumption of fish on Friday no longer compulsory. The fish they contained (mainly carp) provided a good supply of fresh meat throughout the year. While small holding ponds – *servatoria* or 'stews' – were often found near or within the gardens (like the series of four rectangular ponds which still survives to the east of Gedding Hall), the larger ponds, in which the fish were reared and lived most of their lives, were usually located away from the house, often within the park where they could be given more protection from poachers. Such ponds were usually formed by placing earth dams across narrow valleys, like the four shown within Hawstead Park on a map of 1616, made in a valley of a tributary of the river Lark.[55] Other examples still exist at Shrubland (a single pond), Helmingham (two ponds) and Kentwell (an elaborate and complex series). The latter are of particular interest. They lie in an area of woodland about a kilometre to the north-west of the hall, within what was originally the area of the late medieval park. Three main ponds, two of which are still filled with water, are arranged in a flight down a gentle valley, fed by a spring and associated with a number of smaller, auxiliary ponds (Figure 10). The date of this elaborate complex is unknown, but there are references in 1529 to 'the Ponds in the Park', and they were presumably constructed around the same time as the park itself was laid out, in the middle of the fifteenth century.[56] They were still functioning in 1676, when Sir Thomas Darcy sold the hall to Thomas Robinson, together with 'The park stored with above 150 deere, a double dovehouse, fish ponds and other conveniencys, besides timber on the ground and woods considerable'.[57] Warrens for rabbits were also often found within or beside parks, as at Shrubland Hall, Hawstead Hall and Melford Hall, Long Melford.

The social landscape

Both owners and members of their households took great pride in the parks and gardens attached to their residences. James Howell was appointed tutor for a short time to the children of Sir Thomas Savage at Long Melford, and in 1619 wrote to his friend, Daniel Caldwell, a glowing description of the grounds of Melford Hall:

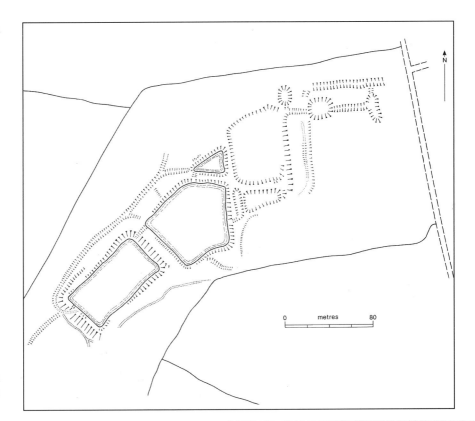

FIGURE 10. The fish ponds, Kentwell.

WITH THANKS TO
KATE BOND

FIGURE 11. The fish ponds, Helmingham Park.

metres
0   80

The stables butt upon the Park, which for a chearful rising ground, for groves and browsing ground for the deer, and for rivulets of water may compare with any for its bigness, in the whole land. It is opposite to the front of the great house, whence from the gallery one may see much of the game when they are a hunting. Now, for the gardening and costly choice flowers, for ponds, for stately large walks, green and gravelly, for orchards, and choice fruits of all sorts, there are few the like in England. There you may have your Bon Christien Pear and Begamott in perfection – your Muscadell grapes in such plentie that ther are bottles of wine sent every yeare to the king ... Truly this house of Long Melford, tho' it be not so greate, yet it is well compacted and contrived with such dainty conveniences every way, that if you saw the landskip of it, you would be mightily taken with it, and it would serve for a choice pattern to build and contrive a house by.[58]

What little information we have suggests, that as in subsequent periods, owners (and particularly their wives) were often closely involved in the creation and maintenance of gardens and took particular pride in the production of the rare and the exotic. On 4 October 1618 Lucy, Countess of Bedford, wrote from Moor Park in Hertfordshire to Jane Lady Cornwallis at Brome:

This monthe putts me in minde to intreate the performance of your promise for some of the little white single rose rootes I saw at Brome, and to challenge Mr Bacon's promis for some flowers, if about you there be any extraordinary ones; for I am now very busy furnishing my gardens ...[59]

Her husband, Nathaniel Bacon, was a particularly keen gardener: his monument in Culford church describes him as 'most knowledgeable in the history of plants'. He was also an accomplished painter, and one of his compositions, *Cooksmaid with Still Life of Vegetables and Fruit* (Figure 12), shows a vast range of vegetables and fruit, including marrows, squashes, pumpkins, cucumber and runner beans (all recent introductions from the New World), some of which he may well have grown himself. Melons also feature prominently, and we know he was cultivating these from a reference in a letter written by Thomas Meautys, his wife's cousin, from London, boasting that should Bacon visit London he would show him 'melons forwarder then his at Broome'.[60] In the background of the picture there appears to be a 'hot bed', in which melons were grown, nurtured by the warmth of decomposing manure – an innovatory practice, recently introduced from the Low Countries. Bacon was not alone in his interest in the cultivation of exotic novelties. Sir William Gage of Hengrave is reputed to have introduced the greengage into England in the early years of the seventeenth century.[61]

The gardens, orchards and other food-producing facilities clustering around the house proclaimed an important, if unsubtle, social message: that the owner ate richer, more varied food than his more lowly neighbours, and was closely involved in the details of its production. Indeed, this was an age in which

costly resources of production, like the elaborate barn which survives at
Seckford Hall, were proudly displayed beside the residences of the gentry.
Dovecotes were a particularly powerful statement of status, as they were
restricted by law to the manorial gentry. The most potent symbol of all,
however, was the park. This usually lay to one side of the house, as at Kentwell,
but sometimes surrounded it on all sides, as at Somerleyton or Hoxne.

It is worth pausing for a while to consider precisely why parks came to be
considered such an important adjunct to a mansion in late medieval times.
Deer and deer parks had long been essential symbols of wealth and status. It
is notable that the upsurge in park-making which occurred in the late fifteenth
and sixteenth centuries coincided with a time not only of economic expansion,
but also of marked upward social mobility – one in which 'new' men were
keen to proclaim their status using that most ancient signifier of wealth and
privilege. As the power of established landed families was challenged by new
wealth, possession of land – and the candid, wasteful display of possession –
thus became a particular mark of distinction. If you've got it, flaunt it –
preferably beside your house. The park moreover had a distinctive 'natural'
appearance which marked out the residence to which it was attached as
something special, distinct from the common working countryside. It was
also a very private landscape – the general population were rigorously excluded

by the encircling pale – and as such served as an arena for exclusive activities, limited to the social elite.

Hunting continued to be the most important of these. It was a highly formalised procedure which involved not only the participants but also a range of spectators. The 1613 map of Melford Hall clearly shows a wooden building raised up on stilts in the south of the park. This was a 'standing', from which the movements of the herds and the hunters could be observed (Plate 4: to the right, is a curious structure, rather like a tree-house, which presumably functioned as a hide for close observation of the deer).[62] But these private, sylvan landscapes were enjoyed in other less formal ways. In 1528 Charles, Duke of Suffolk came with Mary 'The French Queen' to Staverton Park, where they hunted foxes and picnicked under the oaks, 'with fun and games'.[65]

Such recreational activities were not always restricted to deer parks, or to the gardens in the immediate vicinity of the mansion. Some families possessed detached pleasure grounds, located at some distance from both. There was one at Somerleyton in 1652, lying some 1.25 kilometres to the south of the hall, outside the park altogether and separated from it by an area of arable fields. Here, a number of gardens were arranged around the small lake now known (significantly) as Summer House Water, then as Dole Fen (Figure 13). This was apparently in origin a 'Broad', that is, a lake created by the flooding of medieval peat workings. The Somerleyton survey of 1663 describes how:

> The Lady Wentworth hath ... divers fish ponds gardens and walks with a house in farm ... the waters called the Island Pond and the Wall Pond they are outward lying being part in Somerleyton part in Blundeston. There is likewise one other great area of water that is also a private water of the said Lady Wentworth called Dole Fenn with walks and mounts dividing lying part in Somerleyton and part in Blundeston.

The map shows that the Wall Pond lay beyond the western end of, and may have been created by adapting part of, Dole Fen. To its south lay a complicated series of terraces running up the side of the slight valley; to its north was an area of enclosed garden, entered through elaborate gates. To the west of all this, a more substantial area of water – the Hand Pond – is depicted, with an island, on which stands a small building. To the east of the 'Wall Pond' the long, thin waters of Dole Fen stretched away eastwards. The broad was bounded on the north by two further areas of garden – the Vineyard, and the Harp Island – and by the substantial mount referred to in the survey. A small island, again containing a building (this time an elaborate structure with several storeys) stood in the centre of the broad. Traces of these elaborate gardens still survive as low earthworks in the area around Summer House Water.

The parks and garden of the sixteenth and seventeenth centuries were not simply aesthetic landscapes – pleasant prospects to be viewed from the windows of the mansion. They were used by their owners for a variety of recreations,

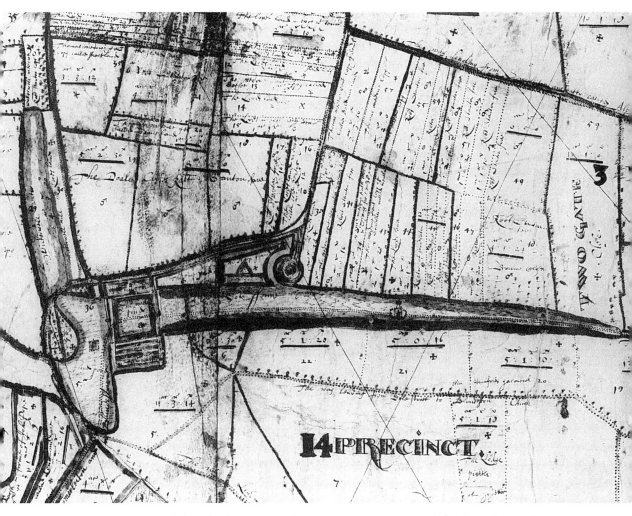

FIGURE 13.
Somerleyton: the gardens at Summer House Water in 1652. Note the mount, and the fishing lodges on islands.

and they had a range of important economic and food-producing functions. Yet at the same time, they were symbolic landscapes, which proclaimed and reinforced the social divisions in a very unequal world.

CHAPTER 3

The development of landscape design, *c.* 1660–1735

Restoration gardens

The Restoration of Charles II in 1660 ushered in a new phase in landscape design in England. Gardens in this period continued to be formal and geometric in appearance, and continued, to a great extent, to be enclosed by walls or fences. But a number of new features appeared. Many landowners had been exiled in Europe during the Civil War and returned to England imbued with the latest French and Dutch ideas of architecture and garden design. The gardens of these countries had developed along different lines during the first half of the seventeenth century, adapting the common heritage of the Renaissance to the particular topographic, social and economic circumstance of each. French gardens were often very extensive and featured complex *parterres* (flowing geometric designs in boxwork) and elaborate water-works. They spilled out into adjacent parks and forests in the form of rides and avenues.[1] These were landscapes appropriate to a country with a low population density and a highly polarised social structure, a nation dominated by a small elite of immensely wealthy individuals – members of the royal family or courtiers. Dutch gardens in contrast were more horticultural in character. 'Cutwork parterres' were employed to display bulbs in patterned flower beds, topiary was a prominent element. They tended to be more compartmentalised than French gardens, that is, they were subdivided inter-nally by lines of trees and hedges. They also made more use of flat areas of water, moats and canals, and were more self-contained: they did not extend out into the surrounding countryside. There were few large parks or extensive forests in Holland, and avenues were not a major feature of landscape design here.[2] All these characteristics, too, can be related to local circumstances. Holland was a young Protestant republic which fed its teaming population through intensive, innovative agriculture, and which had grown rich on long-distance trade. It was controlled by a large mercantile and financial class rather than by a small clique of great landowners. Its level terrain provided few opportunities for elaborate cascades. But at the same time, the low-lying land required assiduous drainage, and garden plants needed to be protected from the winds blowing across the open polders by hedges and tree lines.[3]

Garden historians often discuss the development of English landscape design in this period in terms of distinct 'French' or 'Dutch' styles.[4] But although national circumstances initially encouraged stylistic divergence, by the middle decades of the seventeenth century these distinctions were becoming blurred. French notions of garden design were being widely adopted in Holland and vice versa, the former process encouraged by social and political developments – the growing power of the *Stadtholders* was mirrored in the increasing flamboyance of their gardens. In short, by the time of the Restoration the two styles were becoming confused, even on the continent, so we should not expect to observe pure examples of either in Suffolk or anywhere else in England. Of more importance than such foreign influences in the development of Restoration garden design was the increasing scale of gardens. In a climate of renewed political stability the wealthy had the confidence to invest in their homes, gardens and estates. This was an age of grandiose landscape designs, and ambitious building projects.

The illustration of Brightwell Hall near Ipswich, published by Kip and Knyff in their book *Britannia Illustrata* of 1707, gives a particularly fine impression of a great Suffolk garden of the Restoration period (Figure 14).[5] The grounds of the hall were laid out on a magnificent scale, featuring a number of walled enclosures, lawns, simple *parterres*, and extensive areas of

FIGURE 14. Brightwell Hall, as depicted by Kip and Knyff in *Britannia Illustrata*, 1707.

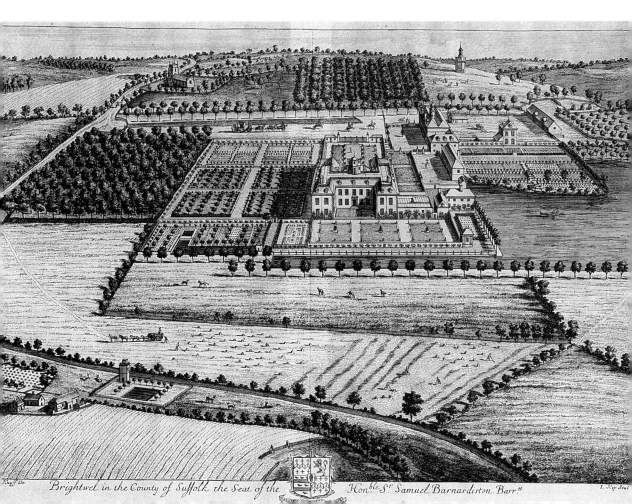

Knyff De.

Brightwel in the County of Suffolk the Seat of the Hon.ble S.r Samuel Barnardiston Barr.tt

I. Kip Scul

water. The hall was demolished at the end of the eighteenth century, but traces of this elaborate layout can be detected in the landscape here today. More sophisticated in many ways were the grounds of Brome Hall, illustrated in the same volume, which featured avenues, wide canals and *parterres* of both grass and gravel with elaborate boxwork and topiary. Here, too, the hall has long gone – its site occupied by a modern house – but more extensive traces survive, including parts of the original walling, a number of terraces and other earthworks, and one of the great canals, although all much modified by later (and especially, Victorian) alterations and additions.

These sites are both, unfortunately, rather poorly documented. But more is known of the magnificent garden laid out at Euston in the last decades of the seventeenth century. Henry Bennet, Lord Arlington, was an important political figure (a member of the CABAL ministry from 1668 to 1672). He bought the manor of Euston early in 1666,[6] and seems to have set about improving the estate immediately: a bill for £23 5s., dated 3 March 1667, survives for work and trees supplied to Lord Arlington at Euston.[7] Phillip Skippon of Foulsham in Norfolk described in his diary for 1669 how he 'saw Euston house yt ye Ld Arlington is repairing and adorning, he hath built very fair stables for about 30 horses, he hath made faire terrase walks in his garden and brought ye river near it by a new cutt ...' – presumably the canal later illustrated by Edmund Prideaux (see figures 22 and 23).[8] The river too was canalised, and the famous arboriculturalist John Evelyn, who visited in 1671 and 1677, tells us that this body of water was terminated at its northern end by a cascade which turned a mill. This not only ground corn but also supplied water to the house and to the fountains in the garden.[9]

Evelyn first visited Euston in October 1671, as a member of the king's party. Arlington asked him for advice on planting his estate which, being on the edge of the Breckland a little to the south of the town of Thetford, was not well-endowed with trees. In Evelyn's own words, 'Here my Lord was pleased to advise with me about the ordering his plantation of firs, elmes, limes etc. up his parke, and in all other places and avenues ...'[10] He was involved in creating the 'wilderness' or ornamental wood, dissected by straight rides, to the south of the hall. Wildernesses were particularly popular in the decades either side of 1700, and a number of other examples are known in Suffolk. The remains of one exist at Easton, to the north-east of the site of the hall, which was demolished in 1925. There are no surviving early maps of the estate, but nineteenth-century surveys describe the area of woodland here as 'The Wilderness' and show that it was crossed by a number of straight rides meeting at a central circular clearing – a typical seventeenth-century arrangement.[11] Another example, laid out around 1710 to the south of Heveningham Hall, again with cross paths meeting at a central clearing, survived until Capability Brown transformed the gardens in the 1780s.[12] One at Loudham Hall, Pettistree, is shown on a map of 1738, occupying the area between the main gardens and the walled kitchen garden (Figure 15).[13] This displays a serpentine pattern of paths, typical of the 1720s and 1730s. Wildernesses had been planted in gardens in the period before the Civil War, as at Somerleyton (above, pp. 18). But their popularity undoubtedly increased in the period after the Restoration, perhaps as a consequence of French influence, and they often came to occupy more prominent positions within gardens, sometimes close to the house.[14]

Avenues were another feature which enjoyed considerable popularity after the Restoration. Evelyn planted a number at Euston. One, composed of limes, ran north–south between hall and church. Others extended the main axes of the house far into the surrounding landscape. The eastern avenue was planted with ash, two rows on each side; the western avenue was of lime, flanking a drive and curving to the south as it approached the hall.[15] Like wildernesses, avenues were not an entirely novel idea – as we have seen, one appears to have been planted at Hengrave as early as 1587 – but their numbers increased massively in the late seventeenth and early eighteenth centuries. Many have disappeared, like those shown on maps of Coldham Hall (1842) or Thornham (1765).[16] However, a number survive in the county, although usually much replanted, generally in the nineteenth century. Noteworthy examples include those at Bradfield Combust (planted *c.* 1725 by Arthur Young's father);[17] Little

FIGURE 15. Loudham Hall, Pettistree, 1739: detail from Kirby's map of 1739, showing the gardens and wilderness.

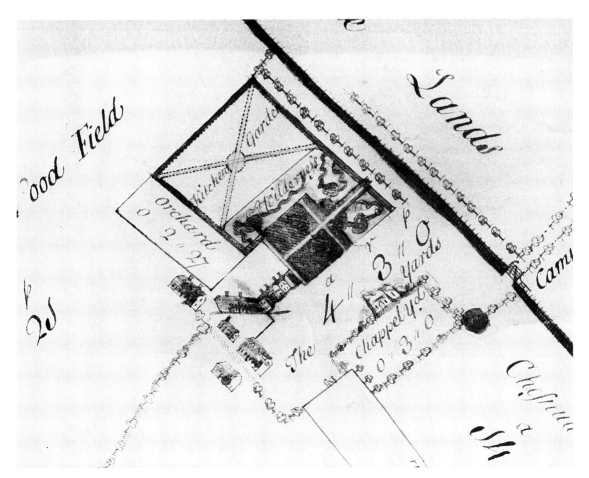

FIGURE 16. The avenue, Little Glemham.

Glemham (Figure 16); Broke Hall, Nacton; Rougham; Somerleyton; and Campsea Ash High House. All are planted with lime, the most popular choice for this purpose in Suffolk, as in most areas of England. Other species were occasionally used, however – as at Heveningham in 1702, where there was a 'Delicate Avenue with 3 Rowes of stately Firr trees upon each side thereof.'[18]

By the early eighteenth century, and probably before, it was common to extend avenues through adjacent areas of woodland as wide rides, as at Dalham (see below, p. 35) or Hintlesham, shown on a map of 1721.[19] Other variations include the creation of formal ponds of water either side of the avenue (as probably at Coldham).[20] Most gentlemen had a single avenue focused on the main façades of the house, but some had a number of alignments, as at Rushbrooke, to the south-east of Bury St Edmunds. A map of 1734, showing the estate's holdings in the parishes of Little Welnethan and Bradfield St George, includes an avenue which ran south-east across the park through an area of woodland to the tower of Bradfield parish church. Unfortunately, the rest of the landscape around the hall does not appear on any surviving map before the Tithe Award of 1843. This and later surveys shows a mesh of avenues and alignments preserved as tree lines and field boundaries to north, south, south-east, and north-west, focused on the hall or its gardens.[21] Another, avenue aligned not on the house but on a point in the park some way to the south, also extended to the north-west. In 1843 this still featured a long linear piece of water, labelled 'The Canal', at its north-western end. Even quite small places sometimes displayed similar 'meshes' of avenues. Great Saxham Hall, for example, had no park as such when surveyed in 1729 – the house stood within a complex of enclosures (including a large and partly moated orchard) close to the village street. But a web of avenues – five in all – extended out across the fields to the south and east.[22] The map details the particular use to which each parcel of land on the estate was put, and it is clear that the avenues all ran across pasture, terminating when they reached public roads or arable fields. It is probable that such complex avenue meshes were fairly common in late seventeenth-century Suffolk. The remains of another, defined partly by avenues running across pasture and partly by rides cut through woods, appears on a map of 1772 showing Barking Hall in mid Suffolk.[23] The same survey shows a particularly long avenue focused on Badley Hall;[24] and another focused on Combs Hall (see below, pp. 44), suggesting that most clayland manor houses of any pretensions had at least one avenue by the early eighteenth century. Given that it was difficult, if not impossible, to maintain lines of trees across fields under arable cultivation, it is likely that many such features disappeared in the late eighteenth century when the clayland pastures were steadily ploughed up.

Avenues are often seen as a manifestation of French influence on English landscape design in the period after the Restoration, and to some extent they probably were. Their popularity may have more complex explanations, however. This was a period in which single-pile houses, often ranged around three sides of a courtyard, were being superseded among the wealthy and fashionable by compact double- or triple-pile houses, more rigorously symmetrical in both external appearance and internal layout. Avenues aligned on the principal façades made a particularly elegant frame for their elegant proportions. They also served to extend the building's main axis of symmetry out into the surrounding countryside, a potent expression of ownership, authority and confidence. In a country which had recently been convulsed by a Civil War, during which the traditional power of landowners had for a time appeared threatened, such things had an obvious appeal and significance.

A strong central axis running out into the adjacent countryside, could be expressed and defined in a number of other ways, as can be seen at Dalham Hall. The hall, which lies close to the parish church and some way to the north of the main concentration of settlement in the parish, was completed around 1705 for Simon Patrick, Bishop of Ely, who purchased the estate in 1702.[25] It is a striking red-brick Queen Anne house of seven bays with a central, three-bay projection; it was originally three storeys in height but was reduced to two following a fire in 1954. A new and imposing landscape appears to have been created when the hall was built. This was organised around a single linear axis ranged roughly north-east/south-west through the centre of the hall. To the south, this axis took the form of an avenue, which is shown on Hodskinson's county map of 1783 (and on the draft Ordnance Survey drawings of *c.* 1815 and the draft enclosure map of 1818).[26] An avenue still exists on this line today, a fine feature composed of horse chestnuts, but this is clearly a comprehensive replanting of the early nineteenth century. Originally, this ran up to and terminated at an east–west public road to the south of the house (which was closed at the time of the Parliamentary Enclosure of Dalham in 1818). On the northern side of the road, a substantial elongated rectangular walled enclosure continued the alignment as far as the house. This enclosure survives, bounded by the churchyard of the parish church to the east and – at least by the start of the nineteenth century and probably originally – by a drive leading up to the stables on the west. It is an impressive feature, apparently coeval with the house. It has piers surmounted with ball finials at its corners and half-way along the length of its long, north–south walls. Its southern side is open, with a drop of *c.* 1.5 metres to outside ground level. This forms a kind of primitive ha ha allowing extensive views back down the avenue and across the Kennet valley – an estate map of 1718 shows that railings originally filled this gap. The map also indicates that this enclosure was originally subdivided by another railing fence half way down its length, into a 'Middle Court' and a 'Nether Court'. It also shows that a number of other enclosures originally lay to the east of the house: a kitchen garden, a 'new kitchen garden' and a bowling green.[27] These, however, were removed in the course of the eighteenth century.[28]

Running northwards, the axis of the landscape passed through the centre of the house, through another walled enclosure (the North Court) and then across a rectangular pond lying some 200 metres to the north of the house (which survives in the parkland). It then continued northwards as a ride cut through a block of pre-existing woodland. A little to the south, the wood was crossed by two further, rather wider rides, intersecting at right angles. These divided it into four compartments – today referred to as Shadowbush, Three Stile, Little Cranes and Big Cranes Woods. Later maps describe this cruciform pattern of rides as 'The Lawns'. The area of woodland may well have originated as a small medieval, or early post-medieval, deer park (the pattern of intersecting rides is strongly reminiscent of the pattern shown on the 1611 map of the parks at Hundon).[29]

In a number of other places in Suffolk significant remains survive of gardens laid out in the years around 1700. Perhaps the most important is at High House, Campsea Ash, 10 kilometres north-east of Woodbridge. The estate was purchased by the Shepherd family around 1648 and remained in their hands until 1883, when it was sold to the Hon W. Lowther. He commissioned Anthony Salvin to extend and partially rebuild the house in fashionable 'Tudor' style. It was completely demolished in 1955, but the gardens and parts of the park remain, beautifully maintained by the present owners.[30] The gardens were probably laid out in the years around 1700, by John Sheppard (1675–1747), following his fortunate marriage to the Dowager Countess of Leicester. A park was created here in the course of the seventeenth century, presumably by the Sheppards; no park is mentioned in the *Chorography* of 1602, but two in close proximity are shown on Kirby's county map of 1735. The pales of the park are also shown on an undated early nineteenth-century map,[31] but otherwise the first details of the landscape are provided by the Campsea Ash Tithe Award map of 1839.[32] At this time an axial avenue led north from the hall. It survives, a magnificent feature, although the present trees (limes with girths in the range of 2.5–3 metres) clearly represent a replanting of the mid-nineteenth century. A little to the west of this feature, and almost parallel with it, the 1839 map shows a shorter stretch of avenue lining part of the main drive to the hall. Only a handful of nineteenth-century horse chestnuts now mark its line, and it remains unclear whether these represent (as at Dalham) the systematic replanting of an original feature or something first created in the nineteenth century. Lastly, a long avenue of elms approached the hall from the east. This survived until the onset of Dutch elm disease in the 1970s. The eye of faith can detect the possible lines of other avenues focused on the house, preserved as lines of field boundaries or alignments of timber, but nothing quite as clear and convincing as at Rushbrooke (above, pp. 34).[33]

The most interesting features, however, survive in the area to the south of the site of the hall. Firstly, there are two magnificent canals (Plate 5). The longest and best preserved is 6 metres wide and 175 metres in length. It has probably been subjected to the attention of nineteenth-century restorers but is impressive nevertheless, with banks which slope down to a narrow berm at the water's edge, and with stone steps running down to the river at various points. It is flanked to the east by a magnificent yew hedge of uncertain age. The other canal, which runs parallel to it and some 75 metres to the west, is now divided by a later causeway and is less well preserved. Between the two canals – and originally occupying the area immediately to the south of the hall – is a grass lawn studded with magnificent ancient cedars, trees which may possibly have begun life as low shrubs growing within some kind of *parterre* (Plate 6). It is very difficult to estimate the age of these trees – their girths range from 5.8 to 7.2 metres – but they were already considered old in the nineteenth century: a sale catalogue of 1883 commented that 'These cedars are believed to be the earliest brought into England (possibly 250 years ago) and here planted.'[34] To the south of the cedar lawn is a kitchen garden. Its walls are partly of late eighteenth- and nineteenth-century date, but the west wall and parts of the southern are probably late seventeenth-century in origin. Perhaps the most intriguing feature of the garden, however, lies to the west of the site of the lost house. This is a circular enclosure defined by massive, ancient yew hedging, bulbous and irregular in outline – aptly described in 1928 as 'strange, picturesque and unique'.[35] It has long been called the 'Bowling Green', and may indeed have served as such – bowling greens were a common feature of gardens of the period – although the area it encloses seems a little small to have served such a purpose. It, too, is probably of late seventeenth-century origin. An illustration in the 1883 sales catalogue shows that, although not as massive or as outgrown as today, it was already a striking feature: 'The celebrated Old Bowling Green ... inclosed by huge and very ancient Yew Hedges, trimmed and cut into fantastic shapes, which are probably the most

unique of their character in England.'[36] Viscount Ullswater, the owner, writing in 1928, was puzzled by the irregular form of growth, the 'strange protuberances', but rightly concluded that they were 'caused probably by slight annual variations in clipping and accentuated by the lapse of time'.[37]

The canals which are such a prominent feature of the garden at High House were present in many other Suffolk gardens created around 1700. That at Abbots Hall, beside the Rural Life Museum in Stowmarket, is particularly striking (Figure 17). Abbots Hall was, in origin, a demesne manor of the Abbey of Bury St Edmunds, and the barn in the adjoining farm complex (now part of the Museum) is an important late medieval structure. The house itself, however, dates from the early eighteenth century (the stack carries a date of 1709).[38] The main feature of the grounds is a large rectangular pond which covers just over 0.7 of a hectare. It is surrounded by a raised terraced walk 2.5 metres wide, still with a crisp outline, especially on the northern side of the pond, where it stands some 1.2 metres above the surrounding land. This is planted on either side with a variety of trees: beech, lime and horse chestnut. Most are comparatively recent but three of the beeches, with girths of *c.* 4 metres, may just possibly represent remnants of the original planting (growing close together they would not put on girth as fast as free-standing parkland trees). Asymmetrically placed towards the western end of the pond is a rectangular island, on which stands a small building that has been variously described as a garden house and a fishing lodge. It is of early eighteenth-century

FIGURE 17. The canal garden, Abbots Hall, Stowmarket.

date, red brick under a red-tile roof, square in plan and measuring 4 metres by 4 metres. It is entered by a door in the western side. The other three sides each originally had windows but that on the eastern side is now blocked. Inside there is fine panelling and a fireplace of *c.* 1800. An enclosure to the east of the pond, now called the Moat Garden, is defined by fairly minor water-filled channels and sunk below the level of the surrounding lawn. Its origins are unclear but it may also have been an eighteenth-century garden feature. It is possible that both pond and Moat Garden began as medieval fish ponds associated with the hall.

Canals were clearly a common feature in late seventeenth- or early eighteenth-century Suffolk gardens. Giffords Hall near Stoke by Nayland – a fine early Tudor house ranged around a courtyard entered by a brick gatehouse – has a large walled garden to the east, probably late seventeenth-century in its present form. The remains of what is probably an early canal survive as a rectangular depression running north–south immediately inside its eastern wall. Another example, created by John Hervey, first earl of Bristol, survives (although in an altered form) beside the walled garden in Ickworth Park. Together with the summer house nearby, it was intended as a feature of ornamental gardens laid out around a new house which was never built; following the demolition of the 'Chiefe Mansion house called Ickworth Hall' here in 1710, the family instead moved to nearby Ickworth Lodge.[39] Many other examples of canals have been destroyed completely, or converted at a later date into more 'naturalistic' serpentine areas of water, like the 'New Canal' dug in front of the house at Culford in 1698.[40]

Canals are often described as a 'Dutch' feature, and the large number found in Suffolk could be seen as a consequence of the county's proximity to and strong links with the Low Countries. While this remains a possibility it is also likely that they owed much to indigenous traditions, and their evident local popularity probably developed naturally out of the long-established Suffolk interest in moats and fish ponds (itself in large measure the consequence of the ease of construction afforded by the heavy clay soils that cover so much of the county). Indeed, the Norfolk landowner and writer Roger North suggests some degree of confusion in the contemporary mind, describing moats in 1713 as 'a Delicacy the greatest Epicures in Gardening court, and we hear of it by the name of Canal'.[41] The canal at Abbots Hall, as we have seen, may have developed from a medieval fish pond complex, and it is a moot point whether the water features at such places as Denniston Hall (two rectangular ponds lying outside the moated enclosure) should be classed as fish ponds or canals. Indeed, most ornamental canals and basins in seventeenth- and early eighteenth-century gardens were stocked with fish. North, among others, recommended that fish ponds should be kept in gardens, partly for reasons of security, partly because 'Your Journey to them is Short and easy, and your Eye will often be upon them, which will conduce to their being well kept, and they will be an Ornament to the Walks.'[42] As late as 1798 an advertisement in the *Bury and Norwich Post* for 'a neat elegant and modern built messuage called Stone Lodge' in Sproughton included a reference to 'a canal stocked with fish'.[43]

Fish ponds in gardens continued, as in the period before the Civil War, to be signs of wealth and status, a visible statement of the superior resources of production and the more varied diet enjoyed by a gentleman. Dovecotes, too, maintained their earlier symbolic status, in spite of the fact that their traditional restriction to the gentry was now being eroded. At Kentwell Hall, Long Melford, a particularly fine early eighteenth-century example stands proudly to one side of the main façade, just outside the moat. Indeed, as in earlier times, gentlemen's houses were generally set within a network of enclosures and yards, some functional, some aesthetic, some both. Typical was Hardwick House, shown on a map of 1663, with entrance court, farm yards, Kitchen Close, orchard, Garden Pightle and Dovehouse Close, containing the dovehouse, all clustering in close proximity.[44] Similar in many ways, although depicted on a map made nearly sixty years later, was Hintlesham Hall. Here the house stood on a moated site, again associated with the usual collection of enclosures, laid out in no very regular order: The Wheat Yard, the Wood Yard, the Stone Court, the Cherry Garden, the Flower Garden (immediately in front of the east façade), the Kitchen Garden, the Wood Yard and the Stable Yard, together with barns and (once again) a dovecote.[45]

The gardens of the minor gentry

The majority of the information we have about gardens comes from the estates of the less exalted individuals and prosperous parish gentry. But many great landowners – minor gentry, vicars and rectors, even yeoman farmers – laid out ornamental gardens, often quite elaborate in nature. The parsonage house at Boxford was probably typical. The Glebe Terrier for 1723 describes how:

> On the East it buts upon a Green Walk bounded by a Canal of a great length, well stocked with Fish; and made by the late Incumbent in a place where before there was Nothing but a foul Stinking Ditch … On the west it fronts ye Road, leading from the Church to Neyland; & buts upon the Kitchen Garden, the Little Court Yard, and Parlour Garden, having before them a Large Square Court Yard, with a Gravel Walk in the Middle, and Lime Trees on each Side of it leading to the Road …[46]

A rather similar collection of features is shown on a map of Bardwell Hall, surveyed in 1730. Here the house was set within a number of enclosures. The largest, the Court Yard, covering just over an acre, lay to the front of the house; behind the house was a garden covering just under an acre, flanked to the west and north by a canal, perhaps created by converting an earlier moat. Much smaller enclosures lay beside the house: these are labelled as 'Wood Yard', 'Front Yard', and 'Garden'. Barns and farm yards lay to the south-east.[47] Similar again is the depiction of Great Bealings Hall on an undated map of *c.* 1700. This shows the house surrounded by a number of small courts, containing three substantial canals. Towards the southern edge of this complex – beyond the largest canal, ranged east–west – was a small wilderness.[48] Here, too, the disposition of the canals makes it highly likely that they were created by the partial infilling of a substantial moat.

Boundary Farm, Framsden, is a particularly interesting case. Here, above a prominent valley to the south-east of Debenham, a late sixteenth-century farmhouse was extended and greatly embellished in the seventeenth century by James Whythe III, member of an upwardly mobile yeoman family who was able to style himself 'gentleman' by 1639.[49] A grand stable was built close to the house, and an elaborate summer house was erected in the garden, set into the wall beside the adjacent public road (rather like those built in the previous century at Melford Hall or Seckford Hall). This building no longer survives, but of particular interest is the long canal or basin to the south of the house, probably added by Wythe's grandson, Edward Mann, in the early years of the eighteenth century. It is aligned east–west and is flanked to the south by a wide terrace walk which affords fine views out over the adjacent valley. When this pond was being cleared of silt in 1990 shallow brick 'steps' were discovered at its eastern end, together with the remains of a wooden culvert, perhaps once forming a diminutive cascade. Similar canals, associated with residences of the lesser gentry, are known from a number of other sites in Suffolk, including Westwood Hall in Stonham Parva. As the archaeologist Edward Martin has emphasised, 'there is a very good chance that further canals are still lurking unrecognised in farmyards, disguised as farm ponds.'[50]

Indeed, Boundary Farm may not have been quite as unusual as we might imagine. A number of seventeenth- and early eighteenth-century maps show quite humble farm houses approached by avenues, and other features which were probably at least in part of an aesthetic character. A survey made in 1722 of High House Farm, Kettlebaston, 'belonging to Sm: Beechcrest Esq Occupied by Henry Making', shows the house approached from the south-east by an axial avenue, and with a small area of woodland to the north-west – presumably an orchard, but perhaps doubling as a wilderness. To the south-west of the house farmyards and outbuildings are shown, but beyond this, in a field called Home Lay, there is another avenue, aligned on the house and described on the map as 'The New Walk'.[51]

Late geometric gardens

The fashion for enclosed gardens, formal canals, avenues and extended axes of symmetry continued into the 1720s and beyond. Glemham Hall, Little Glemham, was extensively rebuilt as a large, rather austere brick building some time around 1720, to judge from the dates on two rainwater heads (1717 and 1722). These alterations were carried out by the North family, who acquired the estate from the Glemhams in the previous century and who continued to reside here until 1920. A park at Glemham is listed in the undated, *c.* 1560 list of Suffolk deer parks but this seems to have disappeared soon afterwards: none appears on Saxton's county map of 1575, nor is one listed in the *Chorography* of *c.* 1602. It was probably re-established when the hall was remodelled and is shown on a fine estate map of 1726, together with the enclosed gardens laid out around the hall and the avenues running out across the parkland to north, south and west (Plate 7).[52] A number of fish ponds are shown within the park, including two substantial rectangular ones in the open parkland to the east of the house. To the north-east of the house a detached area of yards and ancillary buildings, probably including stables, appears to be associated with the remains of an earlier moated site. The garden enclosures to the north, east and west of the hall were removed in the course of the eighteenth century, but the walled enclosure to the south survives, although slightly modified (a lower-walled semi-circular slip at the southern end is a later eighteenth-century addition). Its western wall and parts of the northern are apparently of sixteenth-century date, but the others are of early eighteenth-century construction. The avenue to the north of the house was still in existence in the 1790s, when it was probably felled on the advice of the famous designer Humphry Repton, but that to the south – extending southwards from the walled garden – survives, a particularly fine feature.

FIGURE 18. The gardens at Combs, as shown on an estate map of 1741.

COURTESY SUFFOLK RECORD OFFICE

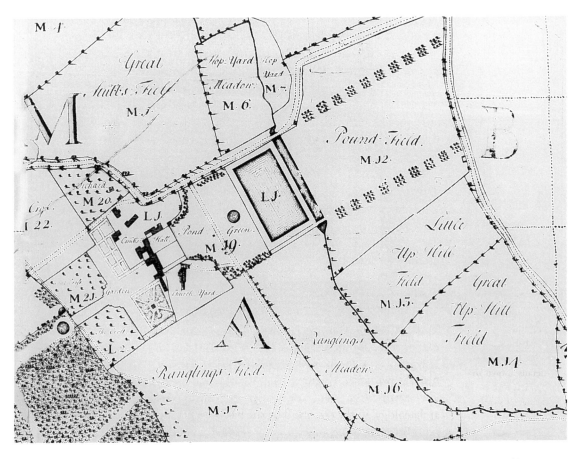

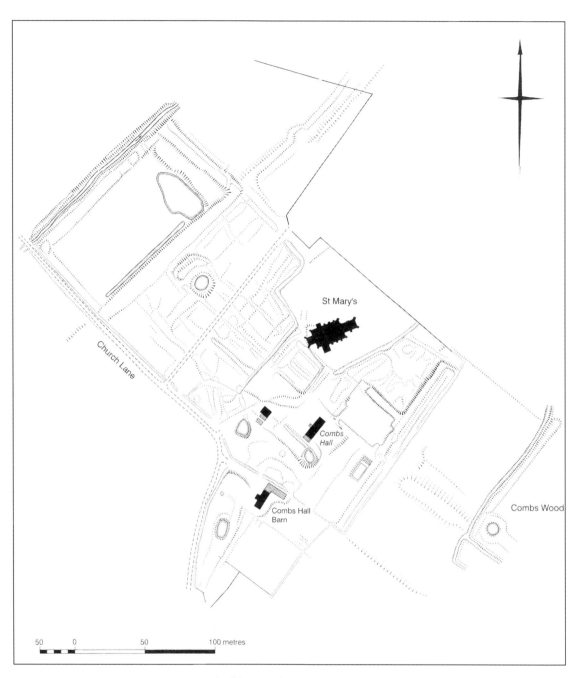

St Mary's

Church Lane

Combs
Hall

Combs Hall
Barn

Combs Wood

50   0   50   100 metres

FIGURE 19. The
earthworks at Combs
Hall (compare with
details shown in
Figure 24).

COURTESY RCHME

It is composed of limes, the oldest with girths of 5 metres or more, probably
representing the original planting of *c.* 1720 (Figure 16).

A rather more ambitious garden was created in the 1720s at Combs Hall
near Stowmarket (Figures 18 and 19). This largely survives in the form of
low earthworks, and constitutes the most striking example of 'garden
archaeology' in the county. The site has been planned by the Royal Commission
on Historical Monuments, and studied by Edward Martin of the Suffolk

Archaeological Unit.[53] The manor of Combs was bought by Sir Edward Gage in 1667, and sold by him twenty years later to William Bridgeman of Westminster, the son of the Amsterdam agent of the East India Company. On his death in 1699 the estate passed to Orlando Bridgeman, an individual with strong Tory sympathies and little political acumen: his will contains a diatribe against 'the fashionable opinions of Deism and as it is called free thinking'. In 1724 Bridgeman built a new brick mansion, slightly to the south-east of the existing, timber-framed manor house, close to Combs parish church.

This house was demolished in 1755 by a subsequent owner, John Crowley, but both it and the associated gardens are clearly shown in a map surveyed in 1741.[54] The landscape was laid out around an axis, nearly a kilometre long, passing through the centre of the house. To the south-east of the house, this was flanked first by *parterres*, and then by groves of trees. Beyond this the gardens ended but the axis was extended as a ride through adjacent woodland, passing over a pond at the wood's edge. To the north-west of the hall, the axis passed through an entrance courtyard, and then across an area of paddock and three ponds: one circular, one large and rectangular, one narrow and rectangular. Beyond, in the direction of Stowmarket, the alignment was maintained by a wide avenue of clumped trees. The fine earthworks in the pasture fields around Combs church confirm the reality of most of the features shown on the map.

Probably of broadly similar date is the garden at Hemingstone, some ten kilometres east-south-east of Combs. Hemingstone Hall was originally built in the mid-1580s and extensively rebuilt around 1625, but the gardens shown on an estate map of 1749 appear to have been laid out around 1720 (Plate 8).[55] The hall sits comfortably at the foot of a fairly steep south-facing slope. The complex pattern of lines and shapes shown on the map evidently represent a network of walks and terraces laid out on this; unfortunately, these were smoothed and levelled later in the eighteenth century, leaving only indistinct traces, but the octagonal feature was probably a platform which both projected forwards from the natural line of the slope, and was also terraced back into it, in the manner of an amphitheatre. It is possible that some of the trees growing here may date back to the time of the garden – including two horse chestnuts with girths of 5.1 and 5.3 metres. To the west of the hall the map shows an enclosure which occupies the site of the present kitchen garden. This has walls which are of various dates: that to the east appears to be of seventeenth-century construction, the others are probably mid-eighteenth century. Marked scarps running east–west through the centre of this enclosure probably represent former terraces, although they do not correspond with any features shown on the map. To the south of the house, the main axis of symmetry was extended, in the customary fashion, by an avenue, which ran out through the fields on the far side of the public road. The map shows that its northern end was flanked by further enclosures. These may well have been nut grounds, for two ancient hazel stools still grow, somewhat incongruously, in the open pasture here.

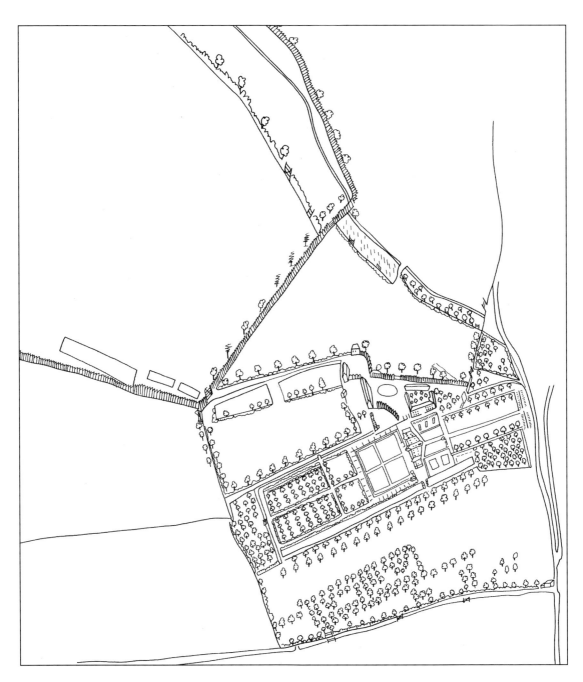

The simple label 'geometric garden' obscures a complex history of development and change. The formal gardens of the early eighteenth century were different in innumerable ways from those created a century before: gentleman's gardens were in a state of constant flux and change. A painting of Thornham Hall (Figure 21), probably made in the 1730s, shows that the lawns and flower borders shown on the early seventeenth-century illustration (above, pp. 10) had been removed, the front wall of the enclosure taken down and the moat

FIGURE 21. The
gardens at Thornham
Hall in the early
eighteenth century,
from an undated
painting.
COURTESY LORD AND LADY
HENNIKER

filled in. The house – now displaying a more rigorously symmetrical elevation – looked out across a simple area of plain lawn, in front of which was a carriage sweep with central statue, separated from the park by elegant iron railings. The remains of the moat had been converted into two rectangular pools, ranged symmetrically to either side of the entrance court. An area of flower garden lay to one side of the house – walled, with cross paths and containing numerous espaliered fruit trees. On the other side was a densely treed area, possibly a wilderness, possibly an orchard, possibly both.[56]

The Thornham painting shows an arrangement of features typical of the 1720s and 1730s. Although the geometric tradition remained strong in Suffolk well into the middle decades of the century, there was a growing fashion for rather simple, elegant gardens, with grass and gravel as prominent features and with flowers often relegated to specialised enclosures to one side of the main façade. Prideaux's illustrations of Euston, again made some time in the 1720s, show the same tendencies on a rather grander scale: a garden of simple, elegant grandeur, with trim lawns, manicured gravel paths, restrained topiary and neatly cut hedges (Figures 22 and 23).[57]

Illustrations and plans of early gardens often pose considerable problems of interpretation, however. In particular, it is easy to assume that the design illustrated is of similar date to the illustration and that it came into existence at a single point in time. In reality, many gardens must already have been old when drawn or surveyed, and must often have been the result of several phases of development. The gardens at Little Thurlow Hall, illustrated on a fine estate map of 1739, are a good example.[58] They occupied a walled rectangular enclosure, approximate 200 by 100 metres, which extended away to the south-west of the hall (Figure 20). A number of compartments were arranged each side of a central axis of symmetry shared by the house: first, what appears to have been a grass parterre or *platt*, divided into four by cross paths and with a banqueting house or summer house built into its the eastern wall;

FIGURE 22 and 23
(*opposite*). The
gardens at Euston
Hall in *c.* 1725.
Edmund Prideaux's
sketches show well the
manicured simplicity
of the late geometric
garden.
COURTESY MR PRIDEAUX-
BRUNE AND NATIONAL
MONUMENTS RECORD

46

next, two open grass areas, surrounded by trees and hedges; and finally two closely-planted wildernesses. The central path then passed through a gate into an area of more irregular woodland. The entire arrangement is very reminiscent of the gardens created at Somerleyton more than a century before (above, pp. 17), but here the main enclosure was flanked (as probably at Campsea Ash) by long canals, in the Dutch manner, so the garden as we see it may well be a composite design which developed over several decades. Elements of this layout, including one of the flanking canals still survive today, although the hall itself was rebuilt in the 1840s.

The use of gardens

We have less information than we would like about what actually went on in gardens in the late seventeenth and early eighteenth centuries. This probably varied considerably, depending in particular on the social status of the owner. Prideaux's drawings of Euston show that the grand gardens laid out around the most important residences were places of parade – arenas of social interaction for the polite, an extension of the formal spaces within the mansion itself. Men like the Duke of Grafton – leading political figures who spent comparatively little time on their estates – probably had little direct involvement with their gardens or, indeed, their home farms. Their pleasure grounds, like their houses, were instruments in the game of politics, ways of displaying the wealth and taste expected of men of power. A little further down the social scale, however, among the local gentry, it is evident that owners and their wives often took a direct and personal interest in their gardens. In the 1680s and 1690s Susanna Betts of Wortham noted down her various gardening activities in her memoranda book, including directions for 'Ordering of my mullions' (melons). The seeds were to be steeped in milk, and sown in a hot bed, 'and when they appear set glasses over them'.[59] Sir Dudley Cullum, who succeeded to Hawstead Place in 1680, was a particularly enthusiastic plantsman. In a long letter sent to John Evelyn in May 1694 he thanked him for his advice concerning the care of tender exotics during the winter: 'and as for my plants, my Orange Trees all winter looked as well as I could wish. My green Ledum … complain'd a little by hanging his leafes, but watering him sparingly brought him to his former verdure …'[60]

Sir Dudley purchased plants from as far afield as London – the nurseryman George Ricketts of Hoxton sent a wide variety of plants and seeds. In April 1687, for example, he dispatched 'the best marigold seed he had', and in May was advising Cullum on how to look after the 'Narcissus of Japan' (although he was unable to supply his customer with a specimen).[61] The Carlisle pear supplied in November 1687 was 'a good bearer', according to Ricketts. Plants were sent by carrier to Bury St Edmunds (4 kilometres from Hawstead) packed in baskets. In one order in 1687 Ricketts sent four cypress trees, a box, a double mirtle, one 'Spannish jessamine', one 'Indian jessamine', and another unnamed plant, in four baskets (the total bill came to £1 10s.).

The extent of Sir Dudley's horticultural activities is demonstrated by the probate inventory drawn up on his death in September 1720. The contents of the greenhouse – a heated masonry building with large windows – were listed in some detail:

One weather glass.
One large Alloe (in a Tubb).
13 Small alloes (in pots).
9 Indian Houseleeks (in pots).
2 Indian Prickley pears (in pots).
24 Orange Trees (in pots).
18 Orange stocks (in pots).
One Cittern and one Lemon tree (in pots).
7 Memomium plennies als Winter Cherries (in Tubbs).
5 Winter Cherries (in tubs).
2 large Bays (in tubs).
One small Bay (in a pot).
2 Mastick plants (in pots).
3 Holianders (in pots).
2 Fremomiums (in pots).
2 Carnegers (in pots).
3 Barba Jovis's (in pots).
4 Marrable Nuts (in pots).
4 Spanish Jessimees (in pots).
2 Leon (in pots).
5 Citises (in pots).
and 18 Murtles (in pots).

In the bowling green were 106 pots of Auriculas and '35 pots of Carnations with 34 sticks and 16 Hoods'. The list of plants growing in the orangery is almost as impressive: it includes aloes, 110 'murtles', 'One horse townge Bay (in a pot)', a pot of periwinkles, five pots of geraniums, three pots of roses, 'One pott of Aetheopia Bramble' and much more.[62]

The reference to a bowling green is a reminder that activities other than horticulture took place in gardens. Bowls was an immensely popular game in the sixteenth and seventeenth centuries, so much so that in 1541 the government, alarmed by the betting it attracted and the neglect of archery it supposedly encouraged, banned it in all places except gardens. Probably most Suffolk gardens of any size boasted a bowling green in the late seventeenth century: examples are known or suspected at Campsea Ash, Little Thurlow (1715), Mildenhall (1720s), Dalham (1718) and Hengrave (1742).[63] The game gradually went out of fashion during the first few decades of the eighteenth century. Sir Thomas Cullum, writing in 1822, suggested that 'Sir Thomas Hanmer, the Speaker, who died in 1746, had a very fine bowling green, contiguous to his house at Mildenhall; and was perhaps one of the last gentlemen of any fashion in the county, that amused themselves with that diversion.'[64]

Many other forms of recreation took place in gardens. In particular, banqueting houses, although perhaps used less frequently than before, were still used for *al fresco* entertainment; old ones were maintained, and some new ones constructed. Sir Thomas Hanmer's banqueting house, probably erected around 1720, survived into the 1820s at Mildenhall. It was 'a summer house of high pretensions ... [and] consisted of one or two rooms on the ground floor, and of a handsome saloon on the first floor.' It was demolished soon afterwards by Charles Bunbury.[65]

The development of parks, *c.* 1660–1730

There was a lull in the creation of new parks during the upheavals of the Civil War and Interregnum, but with the Restoration of Charles II in 1660 park-making began again, with a new intensity. One of the first was created by Lord Arlington around his grand house and fine new gardens at Euston from 1677. Evelyn, as we have seen, gave advice on the planting, and was highly complimentary about the new landscape, describing:[66]

> The Parke Pale which is 9 miles in compas, and the best for riding and meeting the game that I ever saw, with good covert ... In a word this seate is admirably placed for field sports, hunting, and racing.[67]

Although increasingly ornamental in character, parks thus continued to be arenas for a variety of blood-sports, and in particular continued to be stocked with deer. Indeed, in the early eighteenth-century mind the presence of deer defined a park, whatever other functions it might serve and however it might be embellished with avenues and plantations in order to make a pleasing setting for the house. Euston was followed by the parks at Ickworth (*c.* 1704); Glemham (*c.* 1720) and Livermere (by *c.* 1725, but probably expanded in the 1720s and 1730s).[68] By 1735, when John Kirby surveyed his map of the county of Suffolk, there were no fewer than forty-seven parks in the county (Figure 24). Some had originated in the fifteenth and sixteenth centuries, many apparently during the seventeenth, such as Rushbrooke, which was acquired by Sir Robert Davers in 1706[69] and expanded in 1724 when Sir Jermyn Davers added a further 181 acres (*c.* 73 hectares), formerly 'Mr Scots Farm' (an area described in a survey of 1734 as 'The New Park').[70]

The changes in the distribution of parks which had begun in the fifteenth and sixteenth centuries were now more evident. Only a handful of parks could now be found on the level, heavy clays in the 'woodland' district of north-east Suffolk. Most were located on the edge of Breckland or the Sandlings, but there was also a noticeable concentration in the south-west of the county, where the clay plateau is more dissected, the ground more undulating and the soils lighter and more calcareous. Parks continued to be lost as well as created in this period, and while most of the losses were on the heavier clays, the vagaries of descent and the waxing and waning of family fortunes ensured a constant process of change even in the areas favoured by park-makers. Chilton

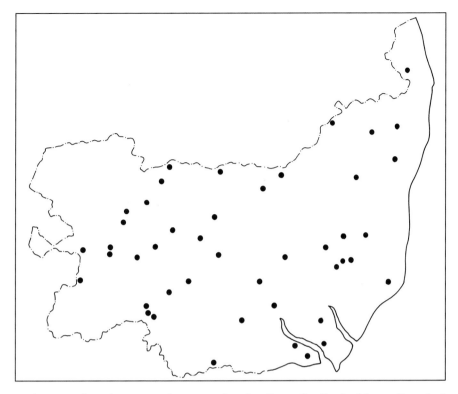

FIGURE 24. The
distribution of parks
in early eighteenth-
century Suffolk, from
John Kirby's map of
Suffolk, 1735.

Park, created in the sixteenth century by the Crane family, had been disparked
before 1715, when a lease referred to 'The two parks being Arable land called
Chilton park by the name of the Hither Park and the Further Park with a
little Lodge standing in the Further Park called the Park House the said
containing by estimation One Hundred and Fifty Acres more or less and
abutting on Cornard Heath on the south ...'[71] The reference to a lodge is a
useful reminder that even post-medieval parks, although located in close
proximity to houses, often contained a lodge to accommodate the park keeper
and his equipment. There are several other examples, including Little Thurlow
park, where a lease of 1706 refers to the 'lodge house in the park'.[72]

Few if any parks could now be found in isolated locations; park and mansion
were now firmly attached. Some lay to one side of the residence, as at Shrubland
Hall, shown on a map of 1688.[73] But an increasing proportion were now laid
out in such a way that the house stood isolated towards the centre of the
park. As manor houses were often located close to other dwellings, this could
have a major impact on settlement, leading in rare cases to the removal of
entire villages. The village of Euston thus appears to have been swept away
when the park here was first laid out. John Evelyn described in 1677 how
Lord Arlington erected a lodge for the park keeper, and 'the same he has done
for the Parson, little deserving it, for his murmuring that my Lord put him
for some time out of his wretched hovell, whilst it was building.'[74] Euston
was a small village before its destruction, but not that small: 11 people were
assessed here by the 1662 Hearth Tax, implying a population of perhaps thirty

or forty.[75] The church was marooned in the pleasure grounds, and was rebuilt in a chaste classical style, now more of a garden building than a spiritual centre for the community.

The removal of Euston was followed, in 1702, by the destruction of many of the farms strung around the greens in the parish of Ickworth, but there were few other instances of this sort of thing in the county. On the whole, Suffolk's settlement pattern is fairly dispersed, featuring a plethora of isolated farms and small hamlets rather than the compact, nucleated villages clustering around church and manor which characterise so much of the Midlands. In the latter region isolated churches within or on the edges of parks are often a good indication that a settlement has been moved. They are a less reliable

FIGURE 25. Loudham Park, Pettistree, surveyed by John Kirby in 1739: a good example of a 'compartmentalised' park surviving into the eighteenth century.
COURTESY SUFFOLK RECORD OFFICE

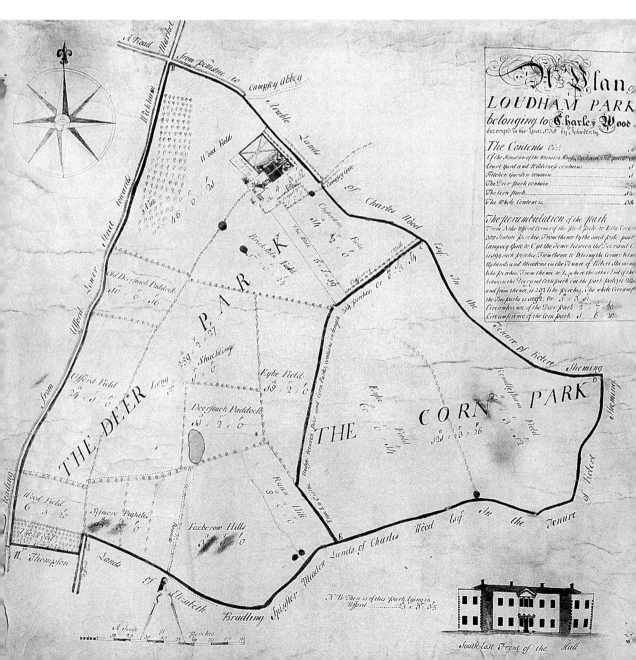

South East Front of the Hall

guide in Suffolk, where farms and cottages often drifted away from old sites beside parish churches to new locations beside greens and commons in the early Middle Ages. We should thus treat with caution the suggestion that the village of Little Glemham was moved when the park was laid out in *c.* 1720.[76] The parish church stands isolated on the southern fringes of the park but there is little indication that a village still clustered around it at the start of the eighteenth century. Equally isolated churches occur in the adjacent parishes of Blaxhall and Farnham, where no parks were ever established.

Although parks and houses were thus generally closely associated, not all Suffolk parks in the early years of the eighteenth century were open expanses of turf, scattered trees and patches of woodland. Some deer parks of archaic, 'compartmentalised' type, containing a number of permanent subdivisions defined by hedges or paling, could still be found. A map of Loudham Hall made in 1739, for example, shows an area called 'Deer Park' divided into a number of hedged sections, each separately named (Figure 25). These were all enclosed by a deer-proof pale which also embraced an adjoining area labelled 'The Corn Park'. The two were divided by 'the hedge between Deer and Corn parks'. The house stood in the north-western corner of the park, surrounded on all sides by yards and gardens.[77] Similar in many ways was the park at Henham, shown on a map of 1699.[78]

The parks and gardens created in the late seventeenth and early eighteenth centuries have left more traces in the modern landscape than those of earlier periods. In part this is simply because they were created more recently, and have thus had less time in which to be obliterated or obscured by later changes. But is also probably because they were laid out on a larger, more durable and more ambitious scale. In the course of the eighteenth century the scale of landscape design was to increase still further, with the steady proliferation of landscape parks.

CHAPTER 4

The triumph of the park

Serpentine gardening

The period between 1730 and 1750 has traditionally been seen as one of crucial significance in the history of English garden and landscape design, in which the formal, geometric style of the sixteenth and seventeenth centuries was gradually simplified, and then made more serpentine and 'naturalistic'. Charles Bridgeman and his contemporaries in the 1730s created landscapes which were still geometric but starker and simpler than those of earlier periods, and in which the distinction between park and garden was often eroded by the replacement of perimeter walls by sunken fences, or ha has (often described as 'fosses' in the 1720s and 1730s).[1] The latter innovation, according to Horace Walpole, writing in the 1760s, was especially important: 'The contiguous ground of the park without … was to be harmonised with the lawn within; and the garden in its turn was to be set free from its prim regularity, that it might assort with the wilder country without.'[2] In the more fashionable gardens, simple grass lawns or 'plats', gravel paths, and wildernesses (now threaded by serpentine paths) became the most important features. In parks, avenues were often now replaced by simpler forms of geometric planting, by rectangular plantations framing vistas towards the house or towards the temples, obelisks and other classical features which were now being introduced into the open parkland.

Under the influence of the designer William Kent, more radical transformations occurred. Areas of irregular, serpentine planting began to appear in gardens, again containing temples and other classical structures. These were idealised representations of the kind of scenery seen in Italy on the Grand Tour. They were also, in effect, three-dimensional versions of the fashionable paintings of such scenery made by artists like Claude or Poussin.[3] In the wider parkland, serpentine watercourses began to make their appearance, and circular clumps of trees became fashionable. In Suffolk, as in other provincial areas of England, these developments were perhaps less marked than they were in the environs of London, but some sites at least were at the cutting edge of fashion.

William Kent was originally a coach-painter from Hull, but his talents as an artist were soon recognised by a group of local gentlemen who paid for him to go on the Grand Tour. While in Italy he met Lord Burlington, who became his principal patron. On his return to England he first worked as an interior designer on the houses of Burlington and his friends, including many of the leading political figures of the time. During the 1720s he became increasingly interested in architecture, and especially in the design of garden buildings. It is not entirely clear where and when Kent first became involved in the design of gardens and landscapes, but there are grounds for believing – as Sally Wilkinson has recently argued – that Euston may be among his earliest works.[4] In 1734 Sir Thomas Robinson described in a now famous letter how a 'new taste in gardening' had arisen 'Which has been practised with so great success at the Prince's garden in Town [i.e., Carlton House in London], that a general alteration of some of the most considerable gardens in the kingdom is begun, after Mr Kent's notion of gardening, *viz* , to lay them out, and work without either line or level.'[5]

Three years before this, however, serpentine features had already been established in the landscape of Euston where, as we have seen, extensive and elaborate gardens had been laid out in the late seventeenth and early eighteenth centuries. The Second Duke of Grafton inherited Euston in 1690 and was appointed Lord Lieutenant of Ireland in 1720, and Lord Chamberlain in 1724. A man of taste and wealth, he began making major alterations to the grounds around 1730. In 1731 the same Sir Thomas Robinson visited Euston, and described the gardens in a letter to his father-in-law, the Earl of Carlisle. The passage is an important one, and worth quoting in full:

> The garden of about 80 acres is fenced on one side from the park by a brick wall in a fosse ... and the slope from the terras in the garden is so wide, that the wall is plante[d] with fruit trees ... On the other side of the fence, between the garden and the park, is a very pretty rivulet cut in a winding and irregular manner, with now and then a little lake etc., and over it in one approach to the house is a wooden bridge built by Lord Burlington, with an arch that appears almost flat and from hence you have a beautiful prospect of the water, which is indeed delightfully disposed. The Park is about nine miles about. The Duke has hitherto done very little to it, but is now entering into a taste, but nature has done so much for him, and his woods and lawns are disposed in so agreeable a manner that a little art and expense will make it a most charming place.[6]

There is no specific reference to Kent here, but the involvement of Burlington in the design of the bridge suggests that Kent may have been involved in the creation of the serpentine watercourse which it crossed: these were one of his hallmarks. Kent was certainly involved at Euston a few years later. In 1738 he wrote to Lord Burlington enquiring, among other things, 'how the mighty works go on at Euston'.[7] By 1743, Horace Walpole was able to suggest that Euston was 'one of the most admired seats in England – in my opinion, because Kent has an absolute disposition of it'.[8] Two drawings by Kent survive at the hall, undated but perhaps made towards the end of the 1730s. One shows a design for a new area of parkland recently added to the west of the house; the other is for a proposed new house in a new location to the west

of the existing site, with a banqueting house on the hill above it (Figure 26). The first of these drawings is rather similar to Kent's design for the North Lawn at Holkham in Norfolk, made around 1740 but only executed a decade later. A classical building – a kind of triumphal arch serving as an entrance lodge – stands in the centre of the picture and is approached by two tracks. The surrounding landscape features wide expanses of turf with scattered clumps of trees which – typically of Kent's work – are actually arranged in a fairly rigidly symmetrical manner, around the central axis leading to the lodge. The latter was eventually built, more or less as shown, and some at least of the planting was carried out, although it is unclear precisely when this occurred. The lodge has a date of 1758 scratched on the roof, but this probably relates to later repairs or alterations.[9]

Kent's other sketch shows a similar mixture of formal geometry and serpentine irregularity. Here the central feature is the Palladian house proposed as a replacement for the existing hall, set in a landscape which is again planted with clumps in almost symmetrical fashion. These contain not only deciduous trees but also numerous conifers – a memory of the landscapes of the Italian *campagna*. On the hill behind the house – sharing its axis of symmetry and

FIGURE 26. William Kent's design for a new house and Banqueting house, Euston, undated *c.*1738.

56

framed by symmetrical blocks of woodland – a large banqueting house is shown. Unlike the proposed new hall, this was, indeed, constructed. Now known as the Temple, it differs only slightly in design from the structure depicted in the drawing, but its position is quite different. It is not placed on an axis running through the middle of the house, but on rising ground some way to the south. The date 1746 inscribed on the inside of the building probably indicates when it was completed. The flanking clumps were planted more or less as proposed, although few of the original trees survive today. The Temple is also shown in a drawing now in the Victoria and Albert Museum. This is signed 'J. Vardy 1755' but was almost certainly by Kent himself, the inscription merely referring to the incorporation of the drawing into Vardy's extensive collection. The building is constructed of local brick, with flint used to give the impression of rusticated quoins.

A rather different landscape, perhaps more typical of its time, was created in the middle decades of the eighteenth century at nearby Culford. Extensive gardens had existed here since the seventeenth century: in 1624 a 'great pond' had been made in front of the house, and in 1698 there are references to a 'new canal'. There was also a park, which was extended around 1714.[10] In 1742 Thomas Wright, a noted designer of gardens and garden buildings, was commissioned to make improvements by the Fifth Lord Cornwallis. His plan shows a landscape of simple geometric form, with a canal and moated water feature to the south of the house, only limited areas of simple *parterres* around it, a short axial avenue focusing on the north elevation and – to the west – a number of geometric plantations, some cut by linear rides and vistas.[11] To the south-east, a larger area of woodland was similarly dissected by straight rides. A striking feature – and one very typical of the period – was the scatter of rather regular clumps in the north part of the park. The status of this plan is unclear – it would appear to be an accurate survey, perhaps made to show alterations recently completed under Wright's direction. Wright certainly paid further visits to Culford in 1744 and 1745, and a sketch of his survives (at Columbia University, New York) of 'A Doric Temple in Culford Park'.[12] It is probable that his main recommendations involved replacing the formal canals to the south of the house with a more fashionable, serpentine body of water, and removing the remaining formal *parterres* around the house, for by 1752 the Dean of Durham Cathedral was able to describe Culford as:

> A sweet place, its scituation clean and dry ... and its park with the advantage of water in front of the house, which extends in a curve for half a mile, but is not seen in front above one third of it, the rest lost among trees. The house stands in a park without any garden around it.[13]

A serpentine water, and an absence of structured gardens across the main façades of the house, indicate that this was a place at the cutting edge of fashion. Nevertheless, serpentine 'naturalism' is in the eye of the beholder: the new watercourse may have appeared very different from the linear and geometric features to be found in most contemporary gardens, but on Repton's plan of the park, made in 1792,[14] the area of water appears rather stiffer and more artificial-looking than the Dean's description suggests, and the park was still dominated by rectilinear plantations.

Similar 'serpentine waters' were appearing in other Suffolk parks by the 1750s. Livermere Park was, as we have seen, created by *c.* 1720 but the adjacent park of Ampton seems to have come into existence only in the 1750s. In September 1753 articles of agreement were drawn up describing how:

> The lands of Baptist Lee Esquire in Great Livermere and Little Livermere have lately been divided from lands of James Calthorpe of Ampton Esquire by a fosse and pale fence and that part of the said land of James Calthorpe Esquire is intended to be covered with water; now Baptist Lee Esquire and James Calthorpe Esquire covenant to erect and maintain a timber bridge over the said water at their equal costs.[15]

The latter was presumably rebuilt before 1765, when a deed of exchange referred to a 'bridge lately built at joint expense'.[16] By the time the first surviving maps of the area were surveyed a few years later, the two parks were indeed separated by a complex water feature.[17] A roughly triangular area of water in Livermere park called the Broad Water was connected by a narrow curving lake – the Long Water – to a broader roughly S-shaped area of water – Ampton Water. The Broad Water was presumably, in large part, a natural feature – the mere from which the villages of Great and Little Livermere took their name. The Long Water and Ampton Water, however, were evidently made by damming an existing natural watercourse – significantly, the parish boundary between Ampton and Livermere continued to run along the middle of these two areas of water.

Brown and his 'Imitators'

The trends apparent at these places – the breaking down of the distinction between garden and park, the removal of all walls and avenues, the adoption of rather simplified geometric planting, in rectilinear blocks and clumps, and the creation of serpentine areas of water – culminated in the decades after 1750 in the work of Lancelot 'Capability' Brown. Brown was the great success story, the very model of upward social mobility, in eighteenth-century England.[18] He was born, the son of a farmer, in 1716 at Kirkhale in Northumberland, and at the age of sixteen took a job as gardener at Kirkhale Hall. In 1739 he moved to the south of England, first to Kiddington in Oxfordshire but soon afterwards to Stowe, Lord Cobham's famous gardens in Buckinghamshire and one of the places where Kent was busy trying out his innovative ideas. By 1741 he was head gardener there, and was soon providing advice for some of his employer's friends. In 1751 he left Cobham's employ, moved to London and set up as an independent landscape designer. His business grew rapidly: Brown supplied not only advice, plans and suggestions, but a whole team of experts (hydrologists, plantsmen) and labourers to carry them out. By the time of his death in 1783 he had worked on more than 170 major commissions, and was by far the most famous landscape designer in England.

It is often said that Brown 'invented' the landscape park, but this is only partly true. Other designers seem to have been creating similar landscapes at the same time, and it is perhaps better to view his style as part of a wider taste and fashion for 'naturalistic' landscapes, which took the ideas of Bridgeman and Kent one step further. All geometry was banished from the park, which now consisted of an open expanse of turf, irregularly scattered with individual trees and clumps. The parkland appeared to run right to the walls of the mansion, although in reality the grazed ground of the park was separated from the mown lawn beside the house by a ha ha, normally more discrete than those in earlier gardens. Farmyards, fish ponds, dovecotes, orchards and the rest were now banished, and the kitchen garden removed or hidden from view. Where finances and topography permitted, a serpentine lake graced the scene, and serpentine rides and carriage-drives meandered through the park, often threading through the woodland belts that ran around much of its perimeter.

Suffolk has less than its fair share of Capability Brown landscapes. There are only five sites where Brown was unquestionably involved: Branches Park at Colinge, Euston, Heveningham, Ickworth and Redgrave. Brown was also consulted about the grounds of Fornham Hall, Fornham St Genevieve, in 1782 by Sir Charles Egleton, but as we shall see the nature of his contribution here remains unclear. Elveden Hall is often suggested as another of Brown's designs. His bank account reveals that between October 1765 and January 1769 he received payments of £1,460 from one 'General Keppell'.[19] It has always been assumed that this was the Hon Admiral Augustus Keppel, who purchased the Elveden estate some time around 1765. It is true that the first reasonably accurate representation of the landscape – provided by the draft Ordnance Survey drawings of *c.*1816–20 – shows a design with broadly 'Brownian' characteristics. In particular, the park was bounded to the west, south and east by a peripheral belt through which ran a serpentine carriage drive, while scattered across the adjacent farmland were a number of clumps and belts, some likewise threaded with serpentine rides. Moreover, on the ground today there are a number of wide, flat-topped circular mounds on the lawn to the south and south-west of the house (one still planted with beech trees) reminiscent of the tree-mounds created by Brown at such sites as Moor Park in Hertfordshire. Nevertheless, the landscape somehow doesn't 'feel' right, and recent documentary research confirms these suspicions. As David Brown has discovered, although Brown received £1,460 from General Keppel, there is no record in our *Admiral* Keppel's account of payments to Brown. The bank account of his relative, General William Keppel of Dyrham in Hertfordshire, *does* however include a payment of precisely the sum in question. We can thus safely remove Elveden from the list of Brown's works in the county – and at the same time add a previously unknown park to the list of Brown's works in Hertfordshire.[20]

Of the five certain Brown commissions in Suffolk, Redgrave – for Rowland Holt – was the earliest. Work began in 1763 and, to judge from Brown's account book, continued on and off until 1773. We often think of Brown simply as a landscape gardener, but he also dabbled in architecture, and although he was initially employed to improve the grounds at Redgrave, his account book makes it clear that he was soon asked to provide a plan for altering the house. This he duly did, encasing its sixteenth- and seventeenth-century core in white Suffolk brick, with stone dressings. The alterations were directed by his associates, Henry Holland and John Hobscroft. The works in the grounds involved, as so often, alterations to an existing deer park: the most important change was the construction of a great lake. Progress does not appear to have been smooth. In 1788 additional expenditure seems to have been required on this part of the project, while in 1777 alterations were made to the shore 'on the London side of the water'.[21] Brown's work included a design for an orangery. The 'temple', or 'round house', which stands above the eastern end of the lake appears to be a later addition, but probably not long after Brown. In all, Brown was paid £10,000 by Holt, the bulk of which was passed immediately to Holland and Hobcroft. In 1771 William Hervey described the water here as 'very fine' and referred to a 'double ha ha', probably along the side of the road – the park was belted around much of its perimeter but open to the south, allowing distant views of the house across the lake.[22] The hall was demolished in 1946; the orangery survived a little longer but has now also gone, and the park is largely ploughed, but the lake and 'round house' survive. When the crops are low in the winter and early spring it is still possible to obtain an impression of what this magnificent landscape must once have been like.

More is known about Brown's work at Branches, where he was employed from 1763 to 1765 by Ambrose Dickens, who had inherited the estate in 1747. Brown was paid a total of £1,500 for his work here, a substantial sum, suggesting considerable landscaping activity. The commission was not entirely a happy one. Dickens requested some extra work, which Brown carried out, charging a further £59. For reasons which remain unclear, Dickens refused to pay this sum, and in the ensuing argument Brown is said to have torn the bill to pieces in front of his client.[23] No Brown plan for the site appears to have survived but a map surveyed in 1766 shows the results of his work.[24] The house lay on the west side of a park of about 190 acres (75 hectares). House, pleasure grounds, an area of offices and a kitchen garden all lay within a large oval area which was separated from the park not, apparently, by a ha ha but by a post-and-rail fence (Figure 27). There was no lake as such, although there were several large ponds in the park and these may well have been modified (although probably not created) by Brown. That to the south had, in particular, a markedly serpentine outline. The park was enclosed by narrow belts, and the interior was scattered with trees, very closely grouped in some places. There were also a number of clumps. Nothing is known of the landscape here before Brown's arrival – there was apparently an earlier park, to judge from

FIGURE 27. The park at Branches, Cowlinge, in 1766, showing the result of Capability Brown's recent alterations.

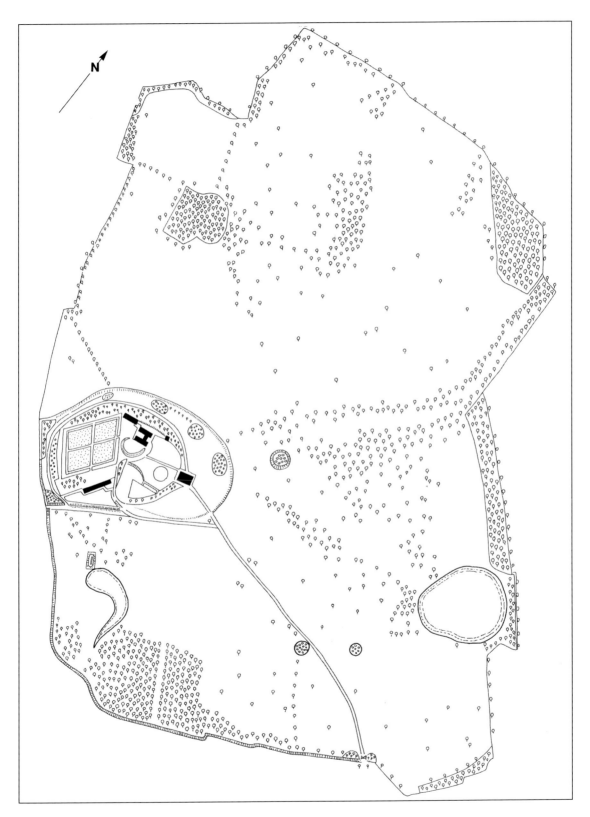

61

John Kirby's county map of 1736, but whether this was very extensive and whether of 'open' or 'compartmentalised' type remains unclear. Certainly, the layout of timber shown on the 1766 map suggests that some was retained from earlier hedgerows, and a particularly marked double line may represent the former course of a roadway approaching the house from the north. Retention of existing timber in this way was, as we shall see, normal practice when parks were created.

Brown's next commission was at Euston, where he was involved from 1767 until 1771. His activities here seem to have principally involved the area around the river to the west and south of the house.[25] As already noted, Kent (or some unknown designer) had effected significant changes here in the 1720s, creating a serpentine river with a series of small lakes. Brown converted these into a single, slightly serpentine body of water. To the south of this, the river was again dammed and some excavations made to create the area of water with a central island now known as The Basin. This was extended to the south as another slightly serpentine lake, called the Broad Water. It is possible that these alterations were, in part, carried out for practical reasons: Joseph Craddock visited the house soon after they had been completed and reported that the Third Duke of Grafton described how 'formerly the great lake was above a foot higher than the floor of the mansion but that now the water appeared to great advantage, yet the whole place was rendered dry and comfortable'.[26] Many of Brown's designs seem to have had as one of their aims the alleviation of drainage problems; the deleterious effects of dampness upon health was a major preoccupation of the age. The records of Drummonds Bank indicate that between 1767 and 1771 Brown was paid £1,997 10s. 6d. by the Duke, a substantial sum, suggesting that these changes involved considerable amounts of earth moving.[27] Doubtless Brown was also responsible for some of the trees and clumps planted around these bodies of water, shown on an estate map of 1772, surveyed by James Parker and a 'Mr Brown'.[28]

Ickworth was Brown's next Suffolk commission, but the character of his work here again remains obscure, in part because of the absence of any detailed eighteenth-century maps of the estate. He was paid £150 by the Second Earl of Bristol in 1769, and there were further payments in 1770, 1773 and 1776, totalling in all £561 8s.[29] This relatively modest sum, spread over a period of seven years, suggests only limited amounts of tree-planting rather than major earth-movement or extensive landscaping (Figure 28). His activities were presumably intended to enhance the setting of the existing house – Ickworth Lodge – where the Herveys were still living at this time (above, pp. 39). The new hall, the present house, was not begun until 1792.[30] Some of the oaks and limes growing in the park today may have been planted by Brown, perhaps to frame views towards and away from the house to Horringer church. The drive leading in from Horringer towards Ickworth Lodge and the area of woodland known as the Grove may also be Brown's work.[31]

When we know that Brown worked on a particular site there is an understandable tendency to assume that everything we see there today of

FIGURE 28. Ickworh Park. Brown made some contribution to the landscape here, but the planting of the park – which was created soon after 1700 by Lord Hervey – probably owes more to the pre-existing countryside.

roughly the right period was the result of his activities. But often he was making alterations to designed landscapes which were already in existence, and which had perhaps only recently been modified by some other designer. Brown, after all, generally worked for the richest clients, men who were accustomed to making frequent alterations to their homes and grounds. Heveningham is a case in point. Sir Gerard Vanneck inherited the estate in 1777 and almost immediately employed the architect Sir Robert Taylor to massively extend and remodel the existing house (built for John Bence around 1714). He commissioned James Wyatt to design the interiors in 1780.[32] Brown's activities here were clearly intended to enhance the setting of this vast new mansion. His account book indicates that he made two visits to the site in 1781, and made two plans: one for the gardens around the house and another for a large lake to the north, a project which would have involved flooding much of the adjoining valley by damming the river Blythe. The same plan shows much new planting of belts and clumps in typical Brown style.[33] The park, which survived in degraded form until recent restoration and alteration, featured two lakes to the north of the house, very 'Brownian' in appearance. Like Redgrave, it was belted around much of its perimeter but open towards the north, so that passers-by on the public road could obtain distant views of the house across the lake.

   Later generations ascribed the design of Heveningham park entirely to Brown, but it is clear that much landscaping had already been carried out before his arrival. A park already existed here in 1752 when Sir Gerard's father, the Dutch merchant Sir Joshua Vanneck, first purchased the estate. He bought other land in the vicinity, including a deer park at nearby Huntingfield. He amalgamated the two parks, and had clearly made a number of 'improvements' before his son inherited.[34] The lakes were not created by Brown. They are shown here on Watt's engraving of the house, based on a watercolour by Thomas Hearne, made some time before 1779 (Figure 29). They must therefore have been created by Sir Joshua. Rather than representing a design for an entirely new era of water, Brown's plan must relate to the expansion and alteration of features already present: a design which was never in the event executed.[35] Several other features in the park sometimes attributed to Brown must likewise have been created before his arrival. Brown's plan shows much planting of clumps and belts in the area to the north of the public road leading to the village of Walpole. But this northerly extension to the park already existed, for it is shown on Hodskinson's county map, published in 1783 but surveyed a few years earlier. Moreover, a description of the area made the following year by a French visitor, Lazowski, suggests planting that was already mature: the area was 'covered by masses of high trees and by clumps very ingeniously placed for good effect'.[36]

FIGURE 29.
Heveningham Hall, as depicted by Watts in *The Seats of the Nobility and Gentry* (1778). Note the boat on the lake and the couple enjoying views from the carriage.

Nevertheless, we should not take these arguments too far. Many alterations were made to the Heveningham landscape as Brown directed. To the south of the house he 'deformalised' existing areas of woodland, which – to judge from Hodskinson's map – were cut with geometric vistas. He also redesigned the pleasure grounds around the house (below, pp. 79); planted new trees and clumps in the open parkland; and laid out a new entrance drive from the north, as well as a number of rides threading in and through the woodland belts. The latter were a classic Brown touch, and were described in some detail by Lazowski:

> There are no walks properly speaking, but there are 16 miles of beautiful grassy ride, turfed so that the open horse-carriage can drive over this superb country. The soil is clayey and rich, and everywhere there are views more or less extensive; everywhere there are certainly fine rich farmlands before your eyes. Sometimes you are under cover of woodlands; always the ride is arranged so that you get the most enjoyable perspectives . . .[37]

As the observant Lazowski realised, Brown's landscapes were not quite so artless and 'undesigned' as they appeared, and carefully contrived views were usually an important component. His plan for Heveningham refers to a 'new portico' which was to be added to the farmhouse to the north-east, and the temple which still exists in the parkland to the south of the house was probably erected around this time.

Brown advised on Fornham St Genevieve in 1782 shortly before his death, and here, too, many 'improvements' had already been carried out. The Fornham estate was owned by Charles Kent, whose great-grandfather, Samuel Kent, a wealthy malt distiller from London, had purchased it in 1731.[38] Some time between 1769 and *c.*1780 the small paddock-like park, avenue and enclosed gardens here were replaced by an extensive landscape park;[39] several public roads were closed; a number of houses and cottages (including a substantial residence a little to the south-west of the hall) were swept away. By the time of Brown's arrival, moreover, the church was in ruins (it was 'consumed by fire on the 24th June 1782, owing to the carelessness of a man who was shooting jackdaws'), and the mill pond in the paddock to the south-west of the house had been extended into a serpentine lake.

Brown sent his surveyor John Spyers to Fornham in March 1782 to make a plan of the site. In September of the same year he himself came down to the estate, accompanied by his foreman Samuel Lapidge, and they subsequently submitted plans for altering the house and rebuilding the church.[40] The need for a survey, however, suggests that landscaping proposals were also being formulated, but it is unclear what, if anything, was carried out. Brown's death would not necessarily have curtailed the execution of any plans, for Lapidge worked as heir to his unfinished commissions. Charles Kent sold the estate in 1788 to B. E. Howard, and a map drawn up in the same year shows a park with a number of typically 'Brownian' features, including scattered clumps and perimeter belt. But whether these were indeed designed by the great man remains unclear; certainly, a road closure order of 1787 indicates further landscaping activity here long after his death.[41]

Although Brown and his works loom large in all histories of eighteenth-century landscape design – and rightly so – other designers, with national practices, were operating at the time, men who languished long in obscurity until rediscovered by recent scholarship. One was Richard Woods, considered a serious rival to Brown in the 1760s and 1770s. His early works (in the late 1750s) were in Berkshire and Buckinghamshire; in the early 1760s most were in Yorkshire or Northumberland; but from 1765 to 1792 well over half were in Essex or the adjacent counties.[42] Only one certain Suffolk commission is so far known, although others may well come to light. In 1777 he was asked by Sir Thomas Gage to improve the grounds of Hengrave Hall (Figure 30). As was often the case, modernisation of a landscape accompanied alterations to the associated residence, and followed hard on the heels of a change in ownership. Gage inherited the estate in 1767 and in 1775 began to make changes to the sixteenth-century hall, changes which involved the demolition

FIGURE 30. The park at Hengrave Hall.

of the outer court and a substantial range of buildings to the east and north, as a result of which the house was reduced to about a third of its original size.[43] A map of 1769 shows that the park still contained a number of old-fashioned, formal features (indeed, there had been little change here since an earlier map, of 1742, had been surveyed).[44] The axial avenue planted in the 1580s survived, complete with terminal viewing mount. There was a circular pond and two subsidiary avenues. Wood's plan involved the removal of all these archaic elements and the creation of a 'naturalistic' landscape of scattered trees.[45] The large fish ponds to the south of the house were to be converted into a small serpentine lake, and the entire park was to be surrounded with a near-continuous belt of trees. In the event, Wood's proposals seem to have been only partially executed: the moat was given a more serpentine form, although it was not extended to quite the extent proposed, and a perimeter belt was established right around the park.[46] But the great axial avenue was allowed to remain (it survives to this day), sitting rather incongruously within the eighteenth-century parkland landscape.

Another major designer of the period – and one who was likewise considered a serious rival to Brown – was Nathaniel Richmond. He was a London nurseryman who had begun his career as one of Brown's foremen, setting himself up as an independent designer in the late 1750s.[47] Most of his commissions were in the Home Counties, especially in Buckinghamshire and Hertfordshire, but he was responsible for a number of landscapes elsewhere, including Saltram in Devon and Beeston in Norfolk. There is circumstantial evidence that Woolverstone Park may have been landscaped by him some time in the 1770s. The site lies in gently undulating countryside some four miles south-east of Ipswich, on the southern banks of the Orwell estuary. A park of some kind had existed at Woolverstone as early as 1725 but this seems to have been improved and extended around 1776 when William Berners – who had purchased the estate in 1773 – employed John Johnson to design a new house, the large mansion, of grey brick and light grey stone that survives today.[48] Humphry Repton, in the 'circular letters addressed to former friends' sent out when he first touted for business in 1788, states that 'Mason, Gilpin, Whately and Gerardin have been of late my breviary – and the works of Kent, Brown and Richmond have been the places of my worship.'[49] The places in question are listed in his account book, opened in June 1788.[50] They include landscapes designed by Kent (Oatlands, Stowe, Holkham), Brown (Blenheim, Holkham, Redgrave) and Richmond (Beeston St Lawrence in Norfolk). But Woolverstone is also included, suggesting that it too was designed by one of these men. The possibility that Richmond was the individual in question is strengthened by the fact that he worked in association with the architect John Johnson at a number of places (Richmond designing the grounds and Johnson the house), including Skreens in Roxwell and Terling (both in Essex).[51]

If Richmond was indeed responsible for laying out the landscape around the house, not all contemporaries were complimentary about it. The French tourist François de la Rochefoucauld described Woolverstone as 'an immense ill-kept park', continuing: 'It consists of a large expanse of enclosed ground artlessly covered with turf and trees, the vistas are neglected. In a word it is as nature wanted it.'[52] The park is first depicted in any detail on the Tithe Award of 1839.[53] This shows that the southern and eastern sections of the park were divided by a fence from the rest; this area was probably reserved for deer. The parish church lay isolated, as now, to the west of the hall. Free-standing trees and small clumps are probably depicted only schematically on the map. Larger areas of woodland are shown to the south-east of the house and along the foreshore of the Orwell estuary (this general pattern of planting is also shown on a late eighteenth-century undated engraving depicting the park from the estuary). Circuitous carriage drives in typical Richmond style led through the park. By this time an obelisk, built in 1793 by Charles Berners as a memorial to his father, had been added to the landscape a short distance from the hall. The park went through many alterations in the nineteenth and twentieth centuries, and not much of the eighteenth-century landscape survives, but some of the oaks and sweet chestnuts in the north of the park towards the river Orwell may have been planted by Richmond. Sweet chestnuts seem to have been one of his favourite parkland trees.

The proliferation of parks

It is an enjoyable exercise identifying which parks were the creations of which famous artists and examining what survives of their work in the landscape of today. However, the vast majority of Suffolk landscape parks were not created by Brown, Richmond or Woods but by local designers or by the owner himself, working in association with his head gardener. Doubtless they were influenced in this by landscapes they had visited, locally or further afield, and it is sometimes possible to see similarities between two landscapes which appear too close to be coincidental. The park at Brettenham, for example, exhibits a number of close parallels with that at Branches, designed by Brown.[54] In both parks, the house is situated on the west side of the park, just south of centre, within an oval enclosure which embraces gardens, yards and service buildings. In both, the layout of these areas – the sites occupied by the various offices and facilities – is very similar, and in both the kitchen garden has a distinctive lozenge-shape. Both parks are liberally scattered with clumps, and both contain a number of large ponds. It is hard to believe that the designer of Brettenham – laid out some time before *c.* 1780 – was not influenced by what he had seen at Branches.

We need to be careful not to compartmentalise garden history too strictly. It is tempting to draw a sharp line at 1750, with the advent of the landscape park and the removal of structure and geometry from the immediate vicinity of the mansion. But in reality the new style was only gradually adopted, and some landowners hung on to their walled gardens in particular for several decades. Moreover, the central importance of the *park* as a setting for a country house was not simply an invention of the 1750s; fashionable sites like Culford and Euston had already, in the 1730s and 1740s, prioritised the park over the garden as the main framework in which to set the mansion. Indeed, the rise to prominence of the park was a long, gradual process which extended back into the Middle Ages, and many of the parks which existed in the 1760s or 1770s were already present by 1700. Some were older still. Nevertheless, parks clearly did increase in numbers during the middle decades of the century, more rapidly than before. The scale of this increase can be quickly assessed by comparing John Kirby's survey of Suffolk, published in 1735, with Hodskinson's county map, published in 1783 but surveyed around 1780 (Figure 31). Kirby's map shows forty-seven parks: these were all, presumably, true deer parks of some kind. Hodskinson's map shows no fewer than seventy-three, an increase of over 55 per cent. There had been losses, as well as gains between these two dates. The parks at Brome, and Brightwell Hall, had both disappeared by 1783.

Some caution is needed in interpreting these sources. Neither of these maps necessarily shows every park in existence: Hodskinson, for example, omits that at Thornham, although this is shown on contemporary estate maps.[55] Moreover, it is not entirely clear what Kirby meant by a park; some of the

FIGURE 31. The
distribution of parks
in Suffolk, *c.* 1780,
from Hodskinson's
county map,
published in 1783.

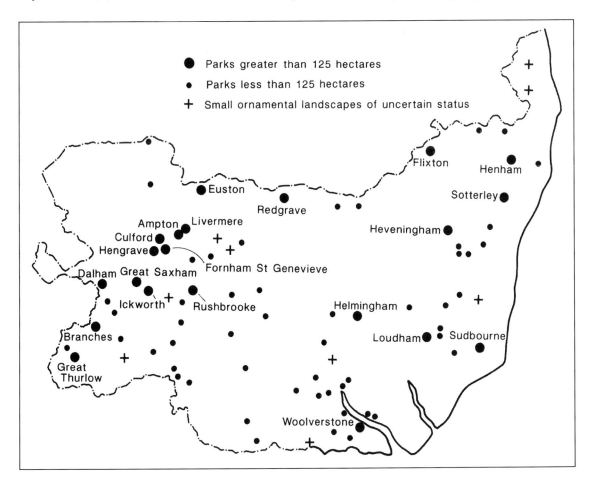

places he included seem, on other evidence, to have been fairly small deer enclosures attached to large mansions. We should also note in passing that Hodskinson may not always be an accurate guide to the size of particular parks; contemporary or near-contemporary maps show that he sometimes exaggerated or included within the park peripheral areas of pasture which they did not consider to be part of the designed landscape.[56] Where parks were relatively small and were not surrounded by prominent belts, and where some of the surrounding land comprised permanent pasture, different cartographers understandably reached different conclusions over what was and was not included within the area of the park. Nevertheless, in spite of these difficulties, the two maps together – Kirby and Hodskinson – probably give a fair impression of the scale of the increase in the number of parks in the course of the eighteenth century.

Unfortunately, it is hard to assess with more accuracy the chronology of park-making across this period. There are, it is true, a number of county maps which were published in the intervening years, but for various reasons these do not appear to be very reliable in their depiction of parks, largely because they were based, to varying extents, on Kirby's original survey. Even that published by his sons in 1766 made little real attempt to update this particular aspect of the map other than by adding the names of current owners – although several parks which had disappeared during the previous thirty years were omitted (such as those at Brome and Brightwell near Ipswich).

It is, moreover, virtually impossible to build up an overall picture of the chronology of park-making from an investigation of individual sites because the history of so many is poorly documented. In most cases, the available evidence merely provides a very approximate date for a park's creation, often a *terminus post quem* or date after which it must have come into existence. Hintlesham Hall, for example, had no park when it was mapped in 1721, but a grant of 1747 refers to 'the park wherein the said capital messuage stands containing 147 acres'.[57] Sotterley Park was probably created following the closure of a number of roads within what became its area in 1746.[58] Orwell Park was created soon after 1757, when John Vernon rebuilt the house here.[59] Drinkstone park probably came into existence around 1760, when Joshua Grigby II, a solicitor from Bury St Edmunds, began to erect a new house here. Plantations at Drinkstone were being vandalised a few years later (below, p. 89):[60] Grigby carried out a large number of land purchases and exchanges here in the 1740s and 1750s, presumably to acquire the land necessary for the large (*c.* 55 hectare) park.[61] Sometimes a *terminus post quem* is provided by some wider documented change in the local landscape. Thus Cavenham Park was created soon after the enclosure of Cavenham in 1773; the area it covered had formerly been occupied by arable open fields.[62] Only occasionally do we have documents which seem to pin-point a park's creation in this period, as at Ampton in 1753.[63] Nevertheless, from these individual cases we obtain an impression – no more – that the rate of park-creation in the county was probably fairly steady through the middle decades of the eighteenth century

– an impression that again emphasises that the advent of Brown did not mark a sudden, decisive break in the evolution of designed landscapes.

Parks which existed before the time of Brown, however, clearly required some modifications to bring them into line with the fashionable norms of the 1760s or 1770s. The later a park was created, the fewer the geometric features it contained, and the less the alterations required to create the required 'naturalistic' landscape. Some early parks probably 'deformalised' to some extent through age and neglect as avenues were felled piecemeal or were brought down by gales and as geometric vistas were gradually softened by loss of trees and spreading branches. At Rushbrooke, for example, age, neglect and the contraction of the park (from 180 to 100 hectares) in the course of the eighteenth century probably contributed as much as any positive actions to the appearance of fragmented, residual geometry apparent on the Tithe Award map of 1843 (above, pp. 34). De La Rochefoucauld described how Sir Charles Davers, its owner,

> Lives a mile and a half from Bury, in a great house which he maintains rather badly, without caring. His park is also neglected: if the grass didn't grow naturally there, there wouldn't be any.[64]

Removal of geometric elements was not the only change that occurred to existing parks in this period, however. Apart from the addition of new clumps, belts and plantations, many expanded in size. At Helmingham, for example, a rental of 1729 describes the park as covering 119 acres (*c.* 48 hectares). By 1770, according to another rental, it covered 351 acres (*c.* 140 hectares). Much of this increase occurred in 1765, when a substantial area to the north of the park – Bocking Hall Farm and North Park Farm – were incorporated into it.[65]

By the time Hodskinson's map was surveyed in 1780 the gradual change in the distribution of parks in the county was complete. Most parks lay, not within the band of heavy clays soils running through the middle of the county, but towards the lighter soils of the Sandlings and Breckland, and the light loams of the south-west – areas where large estates were prominent. The distribution shown by Hodskinson can be analysed further. One noticeable feature is the marked 'packing' of sites in the immediate vicinity of Bury St Edmunds and, to a lesser extent, Ipswich. This is part of a wider pattern. Within the eastern counties as a whole parks tended to congregate noticeably around major urban centres, most notably Norwich and London. Towns and cities were centres of fashionable consumption – places with shops, assemblies and the rest – and the wealthy preferred to dwell within easy reach of them. As early as 1720 Defoe waxed lyrical about Bury, 'a town famed for its pleasant situation, and wholesome air, the Montpelier of Suffolk, and perhaps of England'. It was, in consequence, 'The town of all this part of England, in proportion to its bigness, most thronged with gentry, people of the best fashion, and the most polite conversation'.[66]

Ipswich, too, was a centre for 'polite society': 'There is a great deal of very good company in this town; and although there are not so many gentry here as at Bury, yet there are more here than in any other town in the county.'[67] Other attractions may have contributed to the diminutive cluster of parks around the town: the undulating countryside and fine views across the Orwell estuary doubtless explain the close proximity of Orwell Park, Broke Hall at Nacton, and Woolverstone. Visitors and commentators repeatedly praised the fine views and diverse scenery here. The correspondent in the *Gardeners Chronicle* in the middle of the following century typically enthused about the location of Woolverstone:

> It is impossible to convey by description any true idea of the grandeur of the views: there are five grand openings to the Orwell, which open out the best points on the river nearly the entire distance from Ipswich to Harwich ... The whole of the opposite side of the river ... is nearly an uninterrupted line of wood and park scenery. The fine seat of George Tomline, Esq. of Orwell Park, is right opposite Woolverstone; further down the river is Broke Hall embossomed in sheltering woods and presenting a pretty front towards the river.[68]

The estuary was used, in effect, as a huge serpentine lake, and the views arranged accordingly. Orwell House thus stood at the northern extremity of its park, with the turf – ornamented with a number of neat clumps – stretching away to the shore.[69] Woolverstone, to the south of the river, correspondingly stood some way towards the southern edge of its park.

One further feature of the distribution of eighteenth-century parks in Suffolk, not immediately apparent from the map, is worth emphasising. The cluster of parks down the eastern side of the county forms a rough line, from Henham in the north down to Loudham in the south, placed a little way back from the coast. A quick look at the soil map explains why. Although the estates to which these parks were attached were all concentrated on the light Sandlings soils, the parks themselves lay to one side of this narrow strip of land, on or at least partly on the heavier clay soils to the west. A similar though less pronounced pattern is evident in the west of the county, where parks like Dalham, Great Saxham and Ickworth stand slightly back from the light soils on which the bulk of their estate land lay. The reasons for this pattern are complex and varied. In part, it results from the simple fact that the houses to which the parks were attached were asymmetrically positioned within their estates: they occupied ancient sites chosen – like those of other dwellings and settlements – because of the ease of water supply provided by the perched water table on the clay plateau (water supply was a problem on the dry, porous soils of the adjacent heathlands, except where there were major rivers). But this is only part of the explanation, for some of these houses were established on new sites away from existing settlements in the post-medieval period, and even where they were not, their parks were often laid out asymmetrically, so that the majority of their area could lie on heavier soils.

Eighteenth-century writers on landscape design often talked of the necessity of consulting the 'genius of the place' when improvements were made – the importance of working with, rather than against, the existing landscape. As we saw in Chapter 1, on the light, sandy soils of Breckland and the Sandlings the countryside often remained largely hedgeless and treeless until well into the post-medieval period. These were sheep-corn countrysides, comprising extensive arable open fields and wide tracts of heath. The claylands in contrast had been largely enclosed by the middle of the seventeenth century, and abounded with old hedges, trees and woods, suitable raw materials for the new landscapes.[70] Such areas were naturally more attractive to park-makers than places enclosed more recently from arable land and heath. In particular, ancient hedges could be removed but a proportion of the hedge timber left in place to create an instant sylvan scene. Especially fine examples of such 'pre-park timber' can be seen at Sotterley, where a number of ancient oaks, most formerly managed as pollards, have girths of more than 7 metres. Many are associated with low banks, representing the lines of long-lost hedges or roads removed when the park was laid out (indeed, in an intensively arable county like Suffolk parks often represent the only place in a parish where large numbers of earthworks can be found). Other fine specimens of ancient oak pollards occur in the parks at Helmingham, Ickworth, Little Glemham and Orwell, in all cases with girths of 6, 7 or more metres. Indeed, most Suffolk parks contain some such examples. Pollards are notoriously hard to date; they tend to put on girth at a much slower rate than 'maiden' trees. Old as they look, they are often older still, and some indication of the possible antiquity of these examples is provided by the observation that a line of pollards in Hengrave Park, with girths of between 4 and 5 metres, originally grew in a hedge which is shown on a map of 1588 but which had been removed by 1742.[71] Parks newly created or extended in the nineteenth century also incorporate existing hedgerow trees in this way, as at Benhall, Redisham and Grundisburgh. Moreover, it was not only hedgerow timber that parkmakers used in this way. Many parks, such as Ickworth, Henham and Sotterley, also utilised areas of ancient woodland, often in their perimeter belts, a practice which likewise continued into the nineteenth century, as at Redisham and Great Glemham.

Trees growing in the countryside were thus incorporated wholesale into the new landscape of the park. But much new planting also took place, both of free-standing trees and of clumps and plantations. Its character was perhaps more varied than the surviving trees from this period – overwhelmingly deciduous species, especially oak and (to a much lesser extent) beech – might suggest. Other, less long-lived specimens have disappeared completely over the years. A list of trees planted in the park at Livermere in 1771 thus includes not only 'about 16,000 of Oak, Ash, Elm, Sweet and Horse Chestnut and Beeches' but also 'Norfolk Willows and Turin Poplars' and 'Sicamores'. In 1774, 1,700 'Scotch Firs' – that is, Scots pine – were planted in various plantations in and around the park.[72] Many of these were no doubt used as 'nurses', planted among young deciduous trees to provide protection. But pines were clearly regarded as pleasing in their own right and were sometimes planted out in the open parkland. Some contemporaries were dubious about using fast-growing softwoods in this way; they felt that the slower growing indigenous trees were the mark of the long-established landed family, whilst conifers symbolised a small-profits-quick-returns philosophy which they associated with parvenus from the grubby world of trade and finance.[73] Repton, for example, criticised some of the plantations established at Henham before 1791, asserting that 'a Clump of firs … is only applicable to the recent Villa, and beneath the dignity of an Ancient inheritance.'[74] Nevertheless, larch, pines and poplars – together with sycamores and horse chestnuts – were probably more widely planted in parks and plantations during the eighteenth century than the surviving trees would suggest. Only the hardy cedar of Lebanon, usually planted in pleasure grounds rather than open parkland, survives in any numbers from the eighteenth century (there are fine examples at Heveningham).

We have seen that in Suffolk comparatively few settlements were removed to make way for parks in the early eighteenth century. There are perhaps even fewer examples from the middle and later decades of the century. At Fornham St Genevieve, as we have seen (above, p. 65), several houses and cottages were removed some time between 1769 and 1780 when the park was created. At Great Saxham, a map of 1729 shows the hall standing within enclosed gardens on a public road close to the parish church and three houses or cottages.[75] By 1783, to judge from Hodskinson's county map, these had been cleared and the church lay isolated within a landscape park. There were no doubt other examples, but on the whole, as in earlier periods (and for the same reasons), emparking did not have a major impact on Suffolk's hamlets and villages.

Gardens and pleasure grounds

Garden historians tend to concentrate on the development of the classic landscape park style of Capability Brown and his 'imitators'. Such an emphasis is understandable: the landscape park was a new phenomenon in the eighteenth century, and one which has been described, not without justification, as England's most important contribution to the arts. But such an emphasis tends to imply that gardens disappeared from country houses in this period and that the residences of the wealthy came to stand, solitary and unconnected, in an empty sea of turf. This is a misunderstanding; gardens continued to exist and develop in new ways in this period.[76]

To begin with, we should note that those places like Culford or Euston, where walled gardens were swept away in the 1730s and 1740s to be replaced by open lawns and extended views, remained for a long time the exception rather than the rule. More usual was Ixworth Lodge, where a map of 1740 shows the hall surrounded on all sides by walled gardens in which not only flowers but also vegetables were grown.[77] Mary Lepell, Lady Hervey, described in 1747 how she had been:

for these last three weeks, or indeed a month ... stuck as deeply in my
garden as any of the plants I have set there ... I have made a rosery;
perhaps you will ask what that is; it is a collection of all the sorts of roses
there are, which amounts to fifty ... I have made the whole design of it
myself.[78]

Building accounts for the 1730s and 1740s normally include payments
implying the erection of enclosed gardens. Thus at Tattingstone, the accounts
for 1738–40 include payments for 'Garden walls', 'posts and railes and gate
posts with cutt heads', 'pallisade gates and door' and '2 brass knobs and
latches to garden gates'.[79] Even mansions attached to quite substantial estates
normally retained formal gardens into the 1750s: Hoxne Hall, for example,
had particularly elaborate *parterres* when surveyed in 1757.[80]

Indeed, well in to the 1760s, and in many places into the 1770s, landowners
hung on to their walled gardens, even where extensive parks existed beyond
them. At Thornham, for example, a map of 1765 shows the hall still standing
within the walled gardens shown on the undated painting of *c.* 1720 (Figure
21: above, p. 46). The mansion was approached along an axial avenue which
ran through the open parkland, still passing between the two symmetrically-
placed ponds.[81] More usually, some of the enclosures around the house were
removed, and others retained, so that the owner could savour the delights of
a walled garden *and* enjoy fashionable views across open parkland, as at Dalham
and Little Glemham. At the latter site, interestingly, it was the enclosures on
the northern and western sides of the house, towards the public road, which
were removed and those more discretely located, to the south, which were
allowed to remain. Clearly, many Suffolk landowners would have endorsed
the advice of Horace Walpole:

Whenever a family can purloin a warm and even an old-fashioned garden
from the landscape designed for them by the undertaker of fashion,
without interfering with the picture, they will find satisfactions on those
days that they do not invite strangers to come and see their improvements.[82]

The shelter afforded by walled gardens, especially in the early spring, was
probably an important consideration in a county like Suffolk. But there is
evidence at places like Campsea – where more extensive remnants of geometric
gardens were retained – that some landowners had a general affection for the
old styles of gardening. This can also be seen in the sporadic retention,
throughout the eighteenth century, of avenues (as at Little Glemham, Broke
Hall Nacton, Campsea Ash, or Hengrave).

Archaic features were, not surprisingly, retained longest among those who
lacked the resources to lay out a landscape park of any kind – the minor
parish gentry. The garden of John Sherman at Melton in 1765 featured, not
only walled enclosures, but a long, thin wilderness dissected by straight *allees*,
simple *parterres* and two viewing mounts.[83] Even at the end of the century

some members of this group still had some very old-fashioned things in their grounds. A 'desirable residence and estate' of 240 acres at Theberton, for example, advertised for sale in 1801, had a 'mount and prospect-house, from which are extensive views of the German Ocean, and rich surrounding country'.[84] Nevertheless, retention of enclosures and structured gardens around the house, and even the absence of a fashionable park, did not necessarily indicate relative poverty or a lack of sophistication. There was no park around Hardwick House as late as 1810: farmyards and a walled garden, as well as more serpentine pleasure grounds, clustered around the hall. In 1783 the Rev. Sir John Cullum described to the Royal Society the effects of the recent great frost upon the garden, noting the trees which had suffered most:

> … Larch, Weymouth Pine and Scots Fir had the tips of the leaves withered, the first was particularly damaged. The leaves of some Ashes, very much sheltered in my garden suffered greatly.[85]

He went on to mention 'Walnut, Cherry, Peach, Filbert and Hasel-nut, Barberry, Hypericuym perforatum and hirsutum', as well as a fig tree growing against a west-facing wall.

Such places were becoming exceptional by 1780s, however. By this stage, most landowners with any real pretensions to gentility had removed walled gardens, and contemporary paintings and prints appear to show the house standing within open parkland. Yet even in such cases we should not assume that gardens *per se* had necessarily disappeared. Walled formal gardens may have been removed, but gardens of a kind generally continued to exist. De La Rochefoucauld, visiting Suffolk in 1784, thus described how:

> Near the house, usually immediately round it, is what the English call the garden. It is a small pleasure ground, extremely well-tended, with little well-rolled paths; the grass is cut every week and the trees, which are of rare kinds, grow there naturally, though ever care is taken to prevent moss and ivy growing upon them. In a thousand ways, too, which one does not notice, care is taken to make these gardens attractive. Flowers are planted in them …[86]

At Ickworth Lodge the walled enclosures described above (p. 74) were probably demolished in the 1760s, but in 1770 the Duchess of Northumberland described how the house was 'surrounded by a Garden taken off the Park by a sunk fence and consists of grass dotted all over with Large Trees and a profusion of clumps of flowers'.[87] In a number of places rather more elaborate pleasure grounds were created. Probably the most striking surviving example is at Great Saxham to the west of Bury St Edmunds. In 1774 the old house here, Nutmeg Hall, was demolished and rebuilt by its owner, Hutchinson Mure, with wealth derived from his West Indian sugar plantations. Only five years later this building (partly built to designs by Robert Adam) was destroyed by fire. Mure began to build again but work progressed slowly due to financial difficulties. In 1793 he was declared bankrupt and in August the following year he died.[88] The estate was purchased for £32,000 by Thomas Mills, who redesigned and completed the hall.[89] This stood within a fine park, the design of which was already, by the early nineteenth century, being attributed to Capability Brown.[90] Within the park, in a shallow valley to the east of the house, was a particularly elaborate pleasure ground, threaded with serpentine paths which meandered through woods and shrubberies and past two large pools, the China Pond and the Serpentine Pond, sometimes described as the 'Serpentine River'. De La Rochefoucauld visited in 1784 and noted that 'The garden is agreeable enough, without being very well kept; there is a lake and a river which create a good effect.'[91] To judge from surviving trees, the planting made extensive use of yew and beech. More information about this area is supplied by a survey of the estate made in 1801, which includes elevations of a number of buildings.[92] The 'Temple in the Shrubbery' was the central feature of the pleasure grounds. It survives, now known as the Tea House (Figure 32). Various other buildings were scattered around the estate, including the Temple of Dido, the Temple on the Bridge, and an unnamed structure in another area of pleasure grounds, located to the north-east of the hall. The

FIGURE 32. Great Saxham: the Tea House.

survey shows that the 'Serpentine River' was crossed by a delicate bridge, possibly of ironwork. The gardens continued to develop in the early nineteenth century, when a new building, the Umbrello (a gothic pavilion built of artificial 'Coade' stone), was erected on the far side of the China Pond – perhaps between 1799 and 1813, to judge from the stamp 'Coade and Sealy, Lambeth' to be found at the base of six of its eight columns (figure 33).[93]

Great Saxham is an unusual site in that the main area of pleasure grounds lay detached in the parkland rather than beside the mansion. But there are a number of other known examples in the county. At Helmingham, for example, the obelisk on the low mound some 1.5 kilometres west of the hall marks the site of a summer house, possibly of seventeenth-century date, which stood within a small wilderness/shrubbery.[94] A map of 1803 shows an enclosed area still here, which it describes as 'Pleasure Gardens'[95] (Helmingham was well-supplied with pleasure grounds; another was laid out towards the end of the century by the Sixth Earl of Dysart: – Round Wood, with wide walks and a number of summer houses and statues).[96] Occasionally, pleasure grounds were outside the park altogether. There are few examples in Suffolk; the most striking is Holbrooke, between the estuaries of the Stour and the Orwell, where the Berners family (whose residence was at Woolverstone, 2 kilometres

FIGURE 33. Great Saxham: the 'Umbrello'.

away to the north) had an elaborate garden. David Davy in 1824 gave a detailed description:

> The trees having now grown to a great size, nothing can be more delightful than the appearance of the whole place. A large piece of water is one of the chief and most prominent beauties of it; upon a small promontory at the upper end of it is an elegant building, used as a Summer house, which looks down the water. On the right of this as you pass up to it, a narrow glade conducts you to a smaller piece of water entirely enclosed in wood, except a small knowl on the right hand; nothing can be more retired and romantic than this pond; the trees coming down to the water's edge, and hanging over it in many places; at the lower end on a causeway, artificially raised as dam to the water, is a small summer house. The drive thro' the garden, tho' of no great extent is very pleasing. Some of the trees, particularly a fir or two, standing singly, & being feathered down to the ground, have a very striking appearance.[97]

The origins of this place are unclear. Hodskinson's county map of 1783 marks it as the property of Sir Charles Kent of Fornham St Genevieve on the other side of the county, and he may have intended building a house there, but never did. De La Rochefoucauld in 1784 noted that 'all it wants is a house to make it a very delightful place.'[98] The Berners seem to have acquired it soon afterwards.

Brown, in spite of local tradition, was not responsible for the grounds at Great Saxham. Nevertheless, although he is usually considered the most 'naturalistic' of designers, and the most hostile to structured grounds, he also provided pleasure grounds for his clients. At Heveningham in 1782 he swept away the bastioned wilderness to the south of the house, but gardens were not entirely banished. A ha ha enclosed a large oval area containing the house, a walled garden, an oval lawn surrounded by a path, areas of shrubbery and an orangery, erected a few years later by Samuel Wyatt. There was a flower garden here: de La Rochefoucauld noted that 'Sir Gerard has furnished the slope with a flower garden that is as unsuitable as it is ugly.'[99] Parts of this arrangement – including the ha ha and the orangery – still survive.

It was thus not gardens themselves but walled geometric gardens that became unfashionable in the middle and later decades of the eighteenth century. Moreover, the garden features inherited from earlier decades did not all become equally unfashionable. In particular, there was a fine line between wildernesses and shrubberies, and on occasions even very formal wildernesses – with straight *allees* – might be retained when other elements of geometric landscaping were removed. At Easton, for example, the hall was set in a fashionable pleasure garden separated from the park to the north by a ha ha, but the latter was extended north-eastwards to embrace the earlier wilderness, the essential layout of which – with straight diagonal paths meeting at a central circular clearing – was maintained into the nineteenth century.[100] At its northern corner, at the end of one of the straight *allees*, a substantial summer house was erected on a low mound around 1800, probably replacing an earlier structure. It still survives, but only as a grandiose ruin.

Varieties of park

One commentator described in 1776 how 'The rage for laying out grounds makes every nobleman and gentleman a copier of their neighbours, till every fine place throughout England is comparatively, at least, alike.'[101] And it is true that landscape parks often have a rather stereotyped appearance. But on closer inspection, more variety becomes apparent. A number of broad categories begin to emerge, related to the geographical location of the landscape concerned and to the wealth and social status of its owners. One we might usefully identify comprises large parks (i.e. covering 75 hectares or more) on heavy clay soils. Sotterley in north-east Suffolk is a particularly fine example. It extends over an area of approximately 80 hectares; parts of the park lie on light sandy loams, but most of it occupies heavy boulder clay soils. It is still mostly under grass and is surrounded on all sides by an almost continuous belt of plantations and woods which provide that air of privacy and seclusion so typical of eighteenth-century parks. The house sits centrally within the park, with pleasure grounds to the south and the parish church immediately to the north. The landscape's history effectively begins with Miles Barnes, son of a London merchant and MP for Dunwich, who acquired Sotterley from the Playter family (who had held the estate since the sixteenth century) in 1744.[102] The property then consisted of over 930 acres (*c.* 375 hectares) of land in the Sotterley area, together with another 120 or more in the neighbouring parish of Reydon. Barnes immediately pulled down the existing house and built the present hall on or near its site. Various letters from the 1740s imply that some of the earlier fabric was retained and incorporated into the new building.

The sale particulars of 1744 make it clear that the hall was surrounded by the kind of clutter typical of a country house in this period: 'Stables, Granaries, Walled in Gardens, Kitchen Gardens, Lower Orchards, Court Yards, a Dovecote and land containing about 16 acres'. There was also a brick kiln, a malthouse and 'A broad water with a handsome summer house'.[103] The park probably came into existence around 1746, when writs of *Ad Quod Damnum*, confirmed by royal proclamation in the following year, closed several of the roads running through the area. At the same time there were a number of minor land purchases, including that of a tiny quarter-acre field on which Barnes 'proposed building a summer House ... for a view unto the sea'. In 1746 there are references in the estate correspondence to the construction of a ha ha, and work on 'ye Terras'.[104] The first map to depict the park in any detail, the Sotterley Tithe Award map of 1842, shows a pattern of woods and plantations little different from that which appears on the First Edition Ordnance Survey 6″ of the 1880s and, indeed, survives today.[105]

The 1744 sale particulars boasted that the Sotterley estate consisted not of strips in open fields, but entirely of enclosed land – 'All the Estates above mentioned lye entire and … not one Rood of any Man's land lye intermixed therewith' – a claim confirmed by an undated early eighteenth-century estate map, which shows the property entirely divided into hedged closes of varying sizes. Field names suggest that a deer park had once existed to the north of the hall, but if so it was defunct by the mid-eighteenth century.[106] The woods and the 'many hundreds of timber trees' on the estate referred to in the sale particulars formed the basic planting within the eighteenth-century park, and much of this medieval and early post-medieval fabric survives. As already noted, the park contains a large number of magnificent oaks, mostly former pollards but including some ancient unpollarded standard trees, which once stood in hedgerows (Plate 9). These occur throughout the park, but are particularly prominent towards the south and south-west. Some have girths in excess of 8 metres, suggesting that they must have been planted in the later Middle Ages. Many stand in marked lines or on low banks marking the lines of the former hedgerows. A number of other archaeological relics of the lost clayland countryside are preserved within the park, including a substantial mound, probably a medieval mill mound, and a marked hollow way, the line of a former road, both in the south of the park.

The centre, north and east of the park are occupied by a large area of ancient, semi-natural woodland. Indeed, only the belts running around the southern perimeter of the park appear to have been added when the park was created. Although extensive replanting in the nineteenth century has destroyed most of the coppice stools from within this area, some survive, especially towards the north, where the northern edge of Sotterley Wood is marked by a substantial medieval woodbank. To the south of the woods, on the edge of the open parkland, is a stunning area of massive, outgrown hornbeam coppices, forming tall, multi-stemmed plants with a high canopy. These may well have been incorporated into the park and allowed to grow uncut in this way since the eighteenth century.

The park has many other fine features. There is a small lake in the centre, to the west of the house, with a substantial retaining dam and a canalised stream (with brick retaining wall) to the north. These were apparently created some time between 1746 and 1783 (perhaps through the adaptation of an existing water feature mentioned in the 1744 sale particulars). Beside the south-eastern drive are the remains of a small grotto, comprising a tunnel, 2 metres high and entered through a 'gothic' door, which runs for some 3 metres into a mound. The extensive pleasure grounds to the south of the hall are, in origin, of eighteenth-century date, although most of the current planting is nineteenth- or twentieth-century. Typically, the paths threading through it lead to the kitchen garden – a substantial structure with eighteenth-century walls.

No two designed landscapes are the same, but other large clayland parks display certain similarities with Sotterley, including Redgrave, Flixton, Helmingham, Heveningham and Ickworth. Many have or had lakes or significant water features, extensive areas of woodland and prominent peripheral belts. All make considerable use of existing trees and woodlands. Most, like Sotterley, had medieval deer parks in their vicinity, although only Redgrave developed directly from one; the others nevertheless had comparatively early origins, in the seventeenth or early eighteenth centuries. A survey of Ickworth, made in 1665, shows that the area later occupied by the park was divided into small hedged fields interspersed with woods and areas of wood-pasture – such as the field called Buxall, described as 'a Close of Pasture ... being very well planted with great Tymber'.[107] This agricultural landscape was converted into parkland by John Hervey, First Earl of Bristol, around 1702. The area destined for the park was stoutly fenced, any arable fields were laid to grass, and deer introduced. Some new woods were established, often by planting within existing field boundaries, but as at Sotterley much of the park woodland was of ancient, semi-natural origin.[108] Here, too, large numbers of ancient pollards were incorporated into the new landscape, and many survive to this day, associated with the earthworks of field boundaries and house sites. The park owes more to the ancient, vernacular countryside of wood-pasture Suffolk than it does to the interventions of Brown or any other designer.

Much the same was true of Redgrave, where both the park laid out around the mansion in the fifteenth century, and the surrounding fields eventually incorporated into the landscape park, were generously endowed with trees and woods. A survey made in 1540 described how: 'In the seyd woods and park land about the scytuacons of the seyd mannor and dyvers tenementes there and in other the hamlettes aforseyd and in the lands perteyning in the same be growing 1,100 okes of 60, 80 and 100 years growth parte tymber parte usually cropped and shred'.[109] These trees were incorporated wholesale; William Hervey described Redgrave in 1771 as having 'no timber, all pollards'. There was 'not much fine timber in the park tho' much wood'.[110] Here again Brown's contribution to the landscape was perhaps less important than that made by the pre-existing countryside.

The smaller eighteenth-century parks found in the Suffolk claylands – those extending over 75 hectares or less – display many similarities with those just discussed, but also a number of significant differences. They generally contain numbers of ancient oak pollards and earthwork traces of the pre-existing landscape. But while several originated in the first half of the eighteenth century, few had earlier origins. More importantly, they made less use of pre-existing woodland and were, indeed, generally less wooded than the larger clayland parks. Created for the most part by less wealthy families, parks of this kind lacked the more expensive features; few, in particular, had extensive areas of water. Little Glemham and Thornham are particularly fine surviving examples of parks of this kind.

Different again were those parks established in the heart of Breckland and to some extent within the Sandlings. Sites like Ampton, Livermere, Culford, Euston, Fornham, Tattingstone and Sudbourne had a character quite distinct from that of the clayland parks, although this is not readily apparent as few have survived well: the estates to which they were attached comprised particularly poor soils and thus fared badly during the agricultural depression which set in from the late 1870s. Most have been ploughed up or destroyed by sand or gravel extraction. It is clear, however, that a high proportion of such sites was created in the first half of the eighteenth century, although few originated before the mid-seventeenth century. A significant number were associated with shrunken or deserted settlements – population had often drifted away from villages on these marginal soils from the later Middle Ages, making them prime sites for building mansions and laying out extensive grounds. Most were large – i.e. they covered more than *c.* 75 hectares – and in spite of the dry nature of the local soils, the availability of water in the principal valleys ensured that a number had lakes. Above all, these parks were established in places from which woodland had been cleared in the remote past and which had been enclosed at a relatively late date. They thus made less use of pre-existing trees and woods than the clayland parks, and perhaps rather more use of fast-maturing conifers. The absence of trees and wood, and the bleak and open nature of the surrounding landscape, spurred landowners on to some phenomenal feats of planting, and parks like Lord Cadogan's Santon Downham[111] and Admiral Keppel's Elveden were particularly densely planted with coniferous or mixed coniferous/deciduous clumps which spilled out across the surrounding farmland. These Breckland estates were already being designed with the needs of shooting in mind.

Suffolk's eighteenth-century parks thus displayed a significant, if sometimes subtle, range of variation, largely related to the nature of the local soils, and the character of the existing 'vernacular' landscape. The ancient tradition of the park could, however, be developed in rather different ways. Minor local gentlemen, unable (or perhaps unwilling) to remove walled gardens, farmyards and orchards, and to throw fields together to make a landscape park, often maintained a small area of well-timbered, semi-ornamental pasture beside their house. This was probably an old practice, perhaps with medieval roots. Occasionally, such areas might be dignified with the term 'park' – a map of the estate of John Robinson, in Deniston and Stansfield, dated 1778 shows a 'Park' of ten hectares, while one showing the lands of Reginald Rabett Esq. in Bramfield and Thorington in 1745 has one of less than an acre![112] Such presumption would have attracted ridicule, however, and the more usual term for such areas was 'Lawn' or 'Paddock'. In Suffolk, the latter was normally employed.

Such 'paddocks' feature on many eighteenth- and early nineteenth-century maps, almost always in association with houses still encumbered by enclosed gardens and yards, as at Hardwick Hall in 1810[113] and on the estate of Robert Walpole in Bayton and Hesset, 1795. In the latter case the 'Paddock' covers 14 acres (nearly six hectares) and is quite an elaborate affair, with a number of peripheral clumps.[114] A map of the 'Estate of the Rt. Hon. Lady Dowager Chedworth' in Erwarton, made in 1770, shows the house situated, in the old fashion, between a garden and a farmyard, surrounded by other enclosures, approached by an avenue, and with fish ponds and dovehouse in close proximity (some of these features, including the magnificent sixteenth-century brick entrance gateway and lines of imposing sweet chestnuts, still survive today). To the south-east the 'Paddock', covering around five hectares, is prominently marked, bordered by double lines of trees.[115] Something approaching a definition of the term 'paddock' is provided by sales particulars for a house in Lakenheath in 1829. Lot 4 was 'A PADDOCK of luxuriant Pasture planted with the finest ornamental Timber, at the end of the garden and pleasure grounds, attached to the mansion ...'[116] The term was still widely used even in the second half of the nineteenth century, although now more usually combined with 'park'. Thus at Nether Hall, Pakenham, the ten-acre grass field, 'studded with an abundance of ornamental timber' to the south of the hall was described in sales particulars of 1865 as a 'Park-like Paddock'.[117] It is a moot point where most contemporaries would have drawn the line between a 'paddock' and a 'park' – there was no firm definition, and ideas probably varied from individual to individual, place to place and time to time, but tiny parks like that around the Grove at Yoxford – created in *c.* 1777, expanding to the east in 1785, and attaining an area of 20 hectares by 1839 – or Worlingham, with an initial area, in the 1780s, of *c.* 18 hectares – must have been close to the line.[118]

This chapter has concentrated on the appearance and layout of parks as essentially ornamental landscapes, but deer were still kept in many parks in the later eighteenth century. Indeed, not all the parks shown on Hodskinson's map would have conformed to the Brownian ideal, instead maintaining many aspects of earlier deer enclosures. At Benacre, for example, a survey of 1778, by Thomas Barker of Holton, shows a 'compartmentalised' park which would not have been out of place in the previous century. There were three main subdivisions, and the hall was still associated with yards and service buildings.[119] Interestingly, this old-fashioned park does not appear on Kirby's map of 1735, nor is it mentioned in any other early source. Rather than being a survivor from earlier times, it may have been created following Thomas Gooch's acquisition of the estate in *c.* 1750, perhaps to accompany the new house, designed by the Norwich architect Matthew Brettingham, which was erected around 1764. In a similar way, a map of Loudham Hall in Pettistree, made by Isaac Johnson in 1786, shows that the mansion still lay, as it had done nearly 50 years before, in the far north-western corner of the Deer Park (above, pp. 52–53).[120] The park had now, it is true, lost its internal subdivisions, and the courts to the south-east of the house had been removed so that it faced directly onto the park, but the view here was still framed by a formal avenue; the other façades looked out across gardens, service courts and farmyards to hedged fields beyond. In other words, the park was still, in effect, an old-fashioned deer park rather than a landscape setting for a fashionable house.

The meaning of parkland

Historians disagree over precisely why it became fashionable for large land-owners to surround their house with extensive parks and at the same time remove formal gardens from the main façades. Some have argued that this

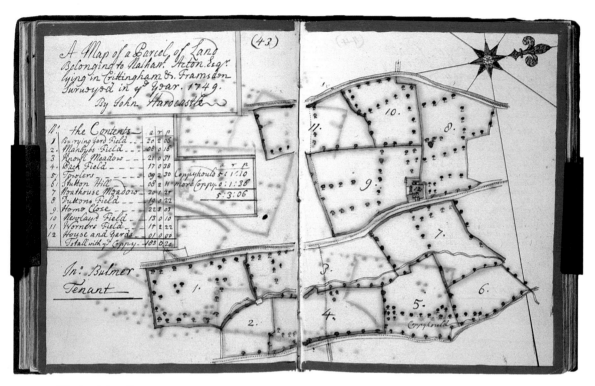

PLATE 1. The 'traditional' clayland landscape: small fields and isolated farms at Crowfield in 1749.

PLATE 2 (*below*). The clayland landscape today: a view near Flixton.

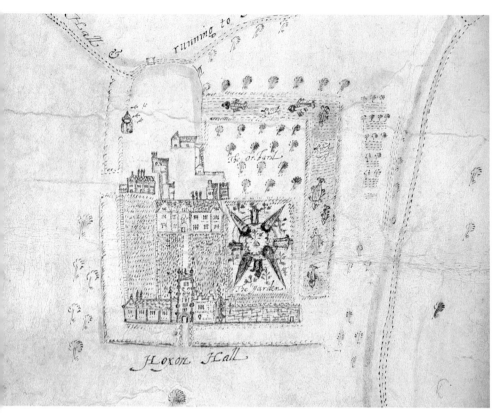

PLATE 3. Detail of a map of Hoxne Hall, 1619, showing the moat and geometric planting.

COURTESY SUFFOLK RECORD OFFICE

PLATE 4. Detail of a map of Long Melford Park, 1613, showing a 'standing' and a hide.

COURTESY SIR JEREMY HYDE PARKER

PLATE 5. High House, Campsea Ash: the canal.

PLATE 6. High House, Campsea Ash: the ancient cedars.

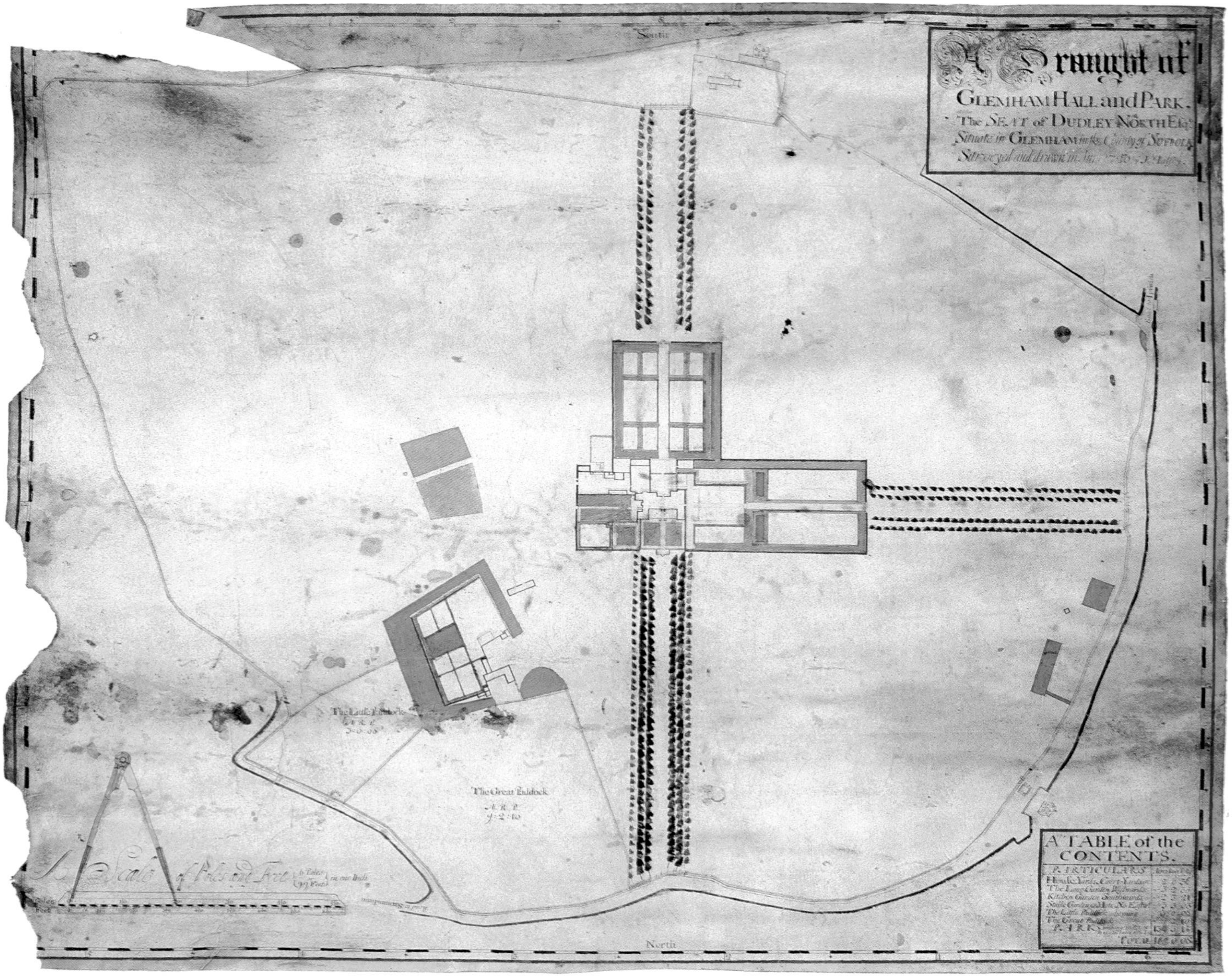

PLATE 7.
Little
Glenham
Hall, gardens
and park, as
depicted on
an estate map
of 1726.
COURTESY
SUFFOLK
RECORD OFFICE

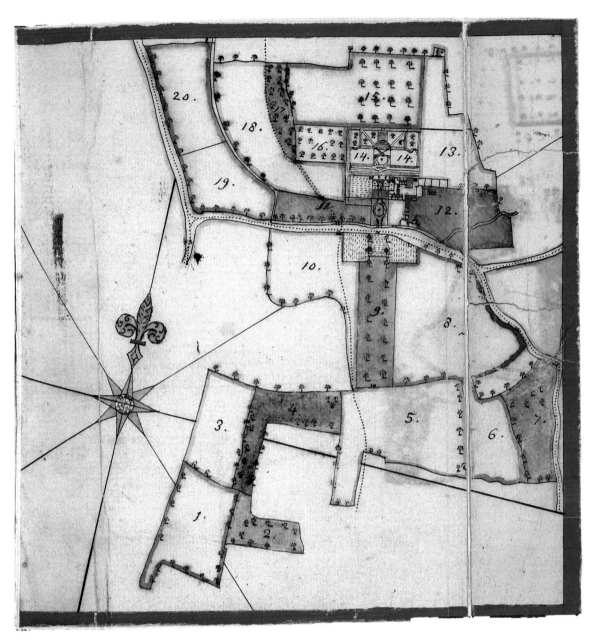

PLATE 8. Hemingstone
Hall: the gardens in 1749.

PLATE 9. Ancient oak
pollards in the park at
Sottlerley.

PLATES 10 and 11. The Red Book for Shrubland Hall, 1789. 'Before' and 'after' views of the prospect from the house. Note the alterations to the prospect tower, and the shrubbery proposed for the foreground.

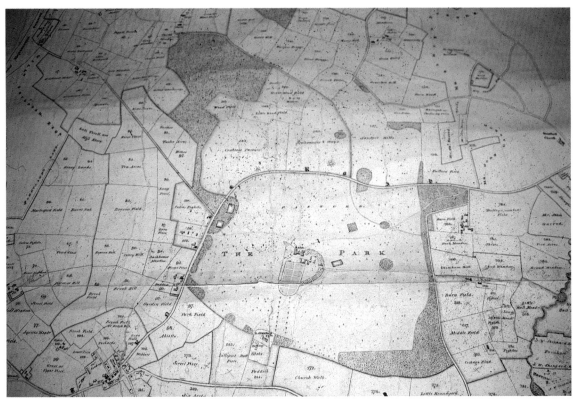

PLATE 12. Little Glenham. The park in 1826, following Repton's improvements. Note the outer belt, and decorated farmland, to the north and wet of the turnpike road.

PLATE 13. Lewis Kennedy, design for Livermere Park, 1815.

PLATE 14 (*inset*). Shrubland Hall: the Descent.

PHOTOGRAPH ALASTAIR TUFFILL

PLATE 15. Shrubland Hall: the Descent, as depicted by E. Adveno Brooke in 1858.

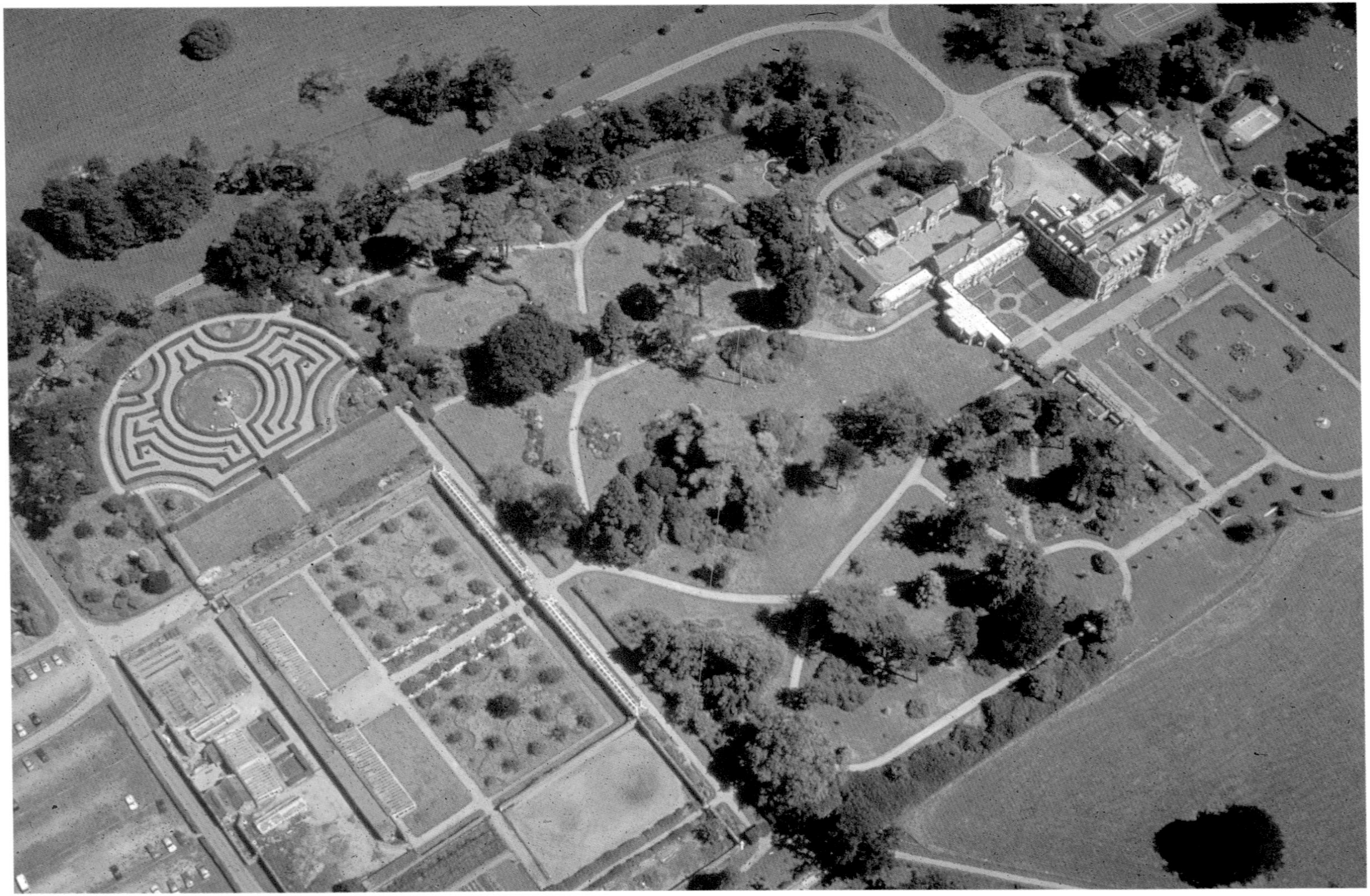

PLATE 16. Somerleyton Hall: aerial view of the gardens. The great *parterre*, now grassed over, occupied the area beside the house on the right of the picture. The pleasure ground of North Lawn is the central area; the kitchen gardens are on the left.

COURTESY DEREK EDWARDS AND NORFOLK MUSEUMS SERVICE

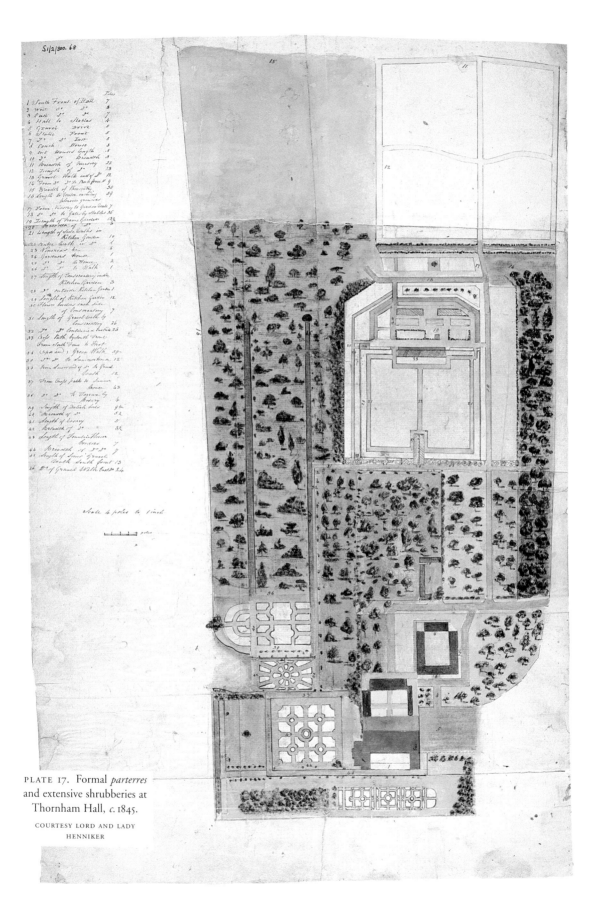

PLATE 17. Formal *parterres* and extensive shrubberies at Thornham Hall, *c.* 1845.

PLATES 18 and 19. The kitchen garden, Great Glemham: one of the best-preserved kitchen gardens in Suffolk.

taste for an extensive 'naturalistic' setting owed much to changes in the wider
landscape. According to this argument, the spread of enclosure was removing
the last traces of wilderness and thus made 'nature' less threatening; this gave
a 'natural' landscape a particular kudos as the setting for a mansion. Moreover,
as the designer John Claudius Loudon put it as early as 1838,

> As the lands devoted to agriculture in England were, sooner than in any
> other country in Europe, generally enclosed with hedges and hedgerow
> trees, so the face of the country in England, sooner than in any other
> part of Europe, produced an appearance which bore a closer resemblance
> to country seats laid out in the geometrical style; and, for this reason, an
> attempt to imitate the irregularity of nature in laying out pleasure grounds
> was made in England … sooner than in any other part of the world …[121]

The problem with such a theory, of course, is that most of England had long
been enclosed by the eighteenth century. Indeed, as should be clear from the
foregoing discussion, most of Suffolk's parks were made at the expense of
anciently-hedged countryside. Changes in aesthetics and philosophy have also
been suggested as factors important in the emergence of the new taste, and
there may be much truth in such arguments. But the Suffolk evidence shows
clearly enough that the landscapes created by 'Capability' Brown and his
contemporaries were in reality the final stage in the long rise to prominence
of the park. It is in this context that we should examine their principal
significance.

The late seventeenth and eighteenth centuries were a period of increasing
polarisation in rural society. Landowners gradually came to associate less and
less with their immediate neighbours, local freeholders or tenant farmers.
Increasingly they sought the exclusive company of their social equals or
superiors. As the eighteenth century progressed, moreover, the numbers of
landless poor steadily increased, and were more and more perceived as a
problem by the wealthy.[122] The rural gentry felt isolated within an uncongenial
world. At the same time, traditional attitudes to social status and traditional
markers of status, were being eroded by what many historians describe as the
'Consumer Revolution' – the phenomenal expansion in the production of
fashionable goods and services which both encouraged, and was in turn fuelled
by, the growth in the size and wealth of the middle classes.[123] A quick look
at the *Bury and Norwich Post* from the middle decades of the eighteenth
century is enough to demonstrate the considerable range of goods and services
now being advertised for sale. These complex social changes had three main
effects on local landowners.

Firstly, they encouraged them to lay out grounds which provided privacy
and seclusion, locating the mansion within an extensive park, closing any
roads and footpaths which ran through it, and obscuring any sight of the less
congenial aspects of the countryside with judicious planting. The closure of
roads was made much easier as a result of legal changes in 1773; few Suffolk
parks had public rights of way leading through them, unless it was a footpath

giving access to an isolated parish church. Parks were not always entirely surrounded by belts of trees – at Heveningham, and Redgrave, for example, they were open to the public road; but here, as in similar cases, the area of intervening turf was so great that a desire for boastful display could be combined with an urge for privacy.

Secondly, these social changes encouraged landowners to emphasise the park as the main setting for the house, and to downgrade the importance of structured, ordered gardens. The reasons for this are not immediately apparent, and require some explanation. There is little doubt that, as part of the general expansion in consumption, middle-class gardens grew steadily in numbers and sophistication. De La Rochefoucauld described in 1784 how Ipswich gave 'the impression of being empty', and suggested that 'this ... derives a little from the spread of the town which is much increased by the large numbers of gardens within its bounds.'[124] Twenty years earlier the Rev. Richard Canning similarly reported of the town how 'most of the better houses, even in the Heart of the Town, have convenient Gardens adjoining them, which make them more airy and healthy, as well as more pleasant and delightful.'[125] The town was already full of elaborate gardens when surveyed by John Ogilby in 1674; these seem more complex and extensive on Pennington's map of 1772.[126] The rural middle class, in small towns and villages, also often possessed elaborate gardens. De La Rochefoucauld, describing the village houses in the area around Woodbridge, noted that 'like all those I have noticed they are neat and brightened by a flower garden.'[127] Advertisements for houses in the local press often emphasised the delights of the associated gardens. The Bury and Norwich Post for 23 May 1798 thus advertised:

> A villa delightfully situated in the parish of Lavenham ... about 10 acres of most excellent pasture land, with pleasure ground laid out with great taste, kitchen garden and melon ground walled, stocked with choice fruits ... fit for reception of a genteel family.[128]

Such evidence as we have, moreover, suggests that gardens at this social level remained highly ordered and structured in appearance (Figure 34). They could hardly be otherwise, given the relatively limited space at the disposal of their owners. Hardly surprising, then, that landowners – keen to proclaim their superior status in this changing world – shunned such gardens, emphasising instead the park which, as well as being a long-established, quintessential symbol of elite status, was also a form of conspicuous display quite unavailable to the majority of the population because it required – and also of course demonstrated – the ownership of broad acres.

Lastly, the growing estrangement of landowners from local communities, and their increasing involvement in fashionable consumption, was accompanied by an understandable decline in their interest in displaying, in the immediate vicinity of their homes, superior resources of production. Orchards, nut grounds, fish ponds, barns and the rest therefore went the way of structured

gardens, and were removed or hidden. At all costs, the residence of gentility must avoid looking like a farm.

Of course, parks themselves had some economic roles, as contemporaries were well aware. But these involved forms of production which had long had a particular association with landowners, especially forestry. The links between tree-planting and gentility were long-established. Small landowners were simply unable to tie up areas of land for long periods as woods or plantations. The clumps, belts and plantations at places like Ickworth [129] were systematically managed for timber. Coppicing generally continued where ancient woods were incorporated into parks, and new areas of coppice might even be established – as at Hengrave, where parts of the west belt, planted in the 1770s, contain hornbeam stools. Trees in parks and plantations were very public investments. When John Hervey came to Ickworth in 1732 he found the park 'disagreeably alter'd' because so many oak trees had been felled by his father, Lord Bristol, because he needed money for 'his multiplicity of children, continuation of heavy taxes, and other accidents'. [130] Such systematic fellings may explain why some parks, even when they have documented histories going back to the sixteenth century, nevertheless have few trees (other than pollards) of any antiquity. At Hengrave, for example – a deer park created in 1587, and systematically landscaped by Richard Woods in the 1770s – the majority of free-standing trees seem to date to the period after 1800, with a significant number having girths in the 3.5–4.0 metre range (implying large scale replanting around 1830).

Expanses of turf, grazed by livestock, were also more closely associated with the social elite than arable fields, which, in all areas of Suffolk, were steadily expanding in the second half of the eighteenth century. Animals symbolised an effortless, patrician form of agricultural production in the way that arable farming – with its large workforce and complex sequence of back-breaking tasks – did not, and many gentlemen took a keen interest in the selective breeding and improvement of livestock. As the Bury St Edmunds land agent John Lawrence noted in 1801, 'There cannot be more interesting objects of view, in a park, then well-chosen flocks and herds, nor one more appropriate to the rural scene, than their voices.'[131] Rare or ornamental breeds were often kept in parks, like the two 'beautiful large horned Holderness Cows pied black and white', advertised for sale in Bury in 1802, 'from a Gentleman's Park in the neighbourhood'.[132] Management of stock of course presented a number of problems. Control of the grazing meant that fences sometimes had to be erected across the open parkland, something which eroded its distinctive appearance, making it appear more like the working agricultural landscape. Sometimes (as at Ampton) sunken fences, or ha has, were employed for this purpose. That at Carlton, in Kelsale, seems to have been made from the hollow way left by a closed road.[133] Parks had to be intensively grazed, not least to provide the close-cropped turf demanded by fashion, and most gentlemen or their stewards found ways of dealing with these matters.

Parks not only had economic functions, however. They were also used by their owners for various forms of 'polite' recreation. In particular, the belts, clumps and plantations were stocked with game, especially pheasants. At Culford, the designer Humphry Repton acknowledged that 'Near a sporting seat the plantations must be thick as cover for game.'[134] One particularly large circular plantation here, he thought, 'might be intermixed with broom and furze to make a strong fox cover'. Indeed, parks played an important role in the social organisation of shooting, as de Rochfoucauld explained during his visit to Suffolk in 1783:

> General custom ... has established a mutual understanding between all those entitled to shoot that a man leaving his own property can go right ahead and shoot anywhere without getting into trouble with the owner provided he doesn't enter the owner's parkland. The rules of polite behaviour forbid this positively.[135]

This was an important consideration at a time when gentlemen were spending more and more money on raising and preserving game birds. In a similar way, lakes were not merely pleasing additions to the prospect. They were used for sailing and fishing. At Culford, Repton advocated a wide, shallow lake rather than one sufficiently deep to allow it to be used for 'pleasure boats'; only the section farthest from the house might be 'made broad and dammed up, to give sufficient depth for fishing boats'. Contemporary illustrations, like that of Heveningham Hall published by William Watts in his volume *The Seats of the Nobility and Gentry* (Figure 29), often show boats sailing on the lake and visitors enjoying the views from the carriage drive. Moreover, although the range of buildings erected at Great Saxham cannot really be paralleled elsewhere in the county, examples are known from several other parks, and most formed destinations for excursions in carriages or on horse back – places to take light refreshments. At Branches, for example, the south-western extension to the park made between 1783 and 1817 was bounded by a plantation which contained, to judge from the sales particular drawn up at the latter date, a tea house.[136] The prospect tower at Shrubland Hall, prominently positioned on the rising ground above the Gipping valley, would have served as a suitable destination for a relaxed ride or carriage drive; the Bacons or the Middletons and their guests could have taken tea there while they enjoyed the fine views.[137]

Eighteenth-century parks were thus playgrounds for the polite as well as private places from which the poor were excluded. Understandably, as rural poverty increased and as resentment towards the rich built up, they sporadically became targets for vandalism. An advertisement in the *Bury and Norwich Post* in 1764 offered a reward of 5 guineas for information about the person who had 'maliciously cut off the leading shoots of several small firs in the plantation made by Mr Joshua Grigsby at Drinkstone'.[138]

CHAPTER 5

Repton and the picturesque

In Suffolk, as in all parts of England, the landscape style of Brown, Richmond, Wood and the rest had triumphed by the last decades of the eighteenth century. Yet even before Brown's death in 1783 some commentators had begun to criticise its blandness and simplicity, and especially the way in which pleasure grounds and gardens were relegated to subordinate or peripheral locations.[1] In 1794 two important works were published, both by Herefordshire land-owners: *The Landscape: a Didactic Poem* by Richard Payne Knight, and *An Essay on the Picturesque* by Uvedale Price.[2] Both writers developed ideas about landscape aesthetics earlier promulgated by the Rev. William Gilpin.[3] They argued that variety and roughness should play a greater role in the design of parks and pleasure grounds. Ideally, such features as blasted oaks, ivy-clad ruins, cliffs and cascades should be employed to add excitement and variety to the scene. In addition, the view from the house should be organised in accordance with the principle of composition adopted by landscape painters like Claude and Poussin; that is, it should incorporate the 'Three marked Divisions' of foreground, middle ground and distance. Brown's landscapes, lacking a foreground and with, in many cases, the far horizon obscured by a belt of trees, clearly failed to pass the picturesque test. The picturesque landscape – as attempted, for example, at Downton Castle in Herefordshire, Richard Payne Knight's own home – had in the far distance hills or mountains, while a foreground might be provided by a terrace or – as Knight suggests in *The Landscape* – by areas of formal gardens in front of the main façade of the house.[4]

It might appear that Suffolk, with its muted terrain and distinct absence of soaring, rugged peaks, was not a place well suited to the creation of picturesque landscapes, and to some extent this is true. But Payne Knight and Price were not only interested in the wilder aspects of nature.[5] Rocky, upland landscapes were greatly admired by Gilpin, Price and Knight, but other kinds of scenery also had a picturesque appeal: forests, for example, or ancient, long-enclosed countryside with knarled old trees, high hedges and deep lanes where:

> The winter torrents, in some places wash down the mould from the upper grounds and form projections … with the most luxuriant vegetation; in other parts they tear the banks into deep hollows, discovering the different strata of earth, and the shaggy roots of trees.[6]

What the prophets of the picturesque prized above all – as this quotation demonstrates – was variety and interest, and this was what they wanted to introduce into the parks and pleasure grounds of the gentry. In particular, they wanted to employ lusher and more varied planting than was usual in eighteenth-century parks. Moreover, while the term 'picturesque' might imply that the landscape should be viewed as a static picture from the windows of the great house, in reality an enthusiasm for variety and surprise ensured that Payne Knight, Price and their followers emphasised the importance of exploration – of carefully contrived paths which served to reveal the hidden beauties of a place. Put simply, designers in the picturesque mode disliked the stereotyped predictability of the Brownian park, as well as the way it was often insensitively imposed on the natural terrain, negating the distinctive character of a particular place.

Repton in Suffolk

By the time Knight and Price published their works, however, another designer had begun to develop a style of landscape which was subtly rather than dramatically different from the sweeping, open parkland of Brown and his 'imitators' – a designer with a particular association with Suffolk. Humphry Repton was born in the town of Bury St Edmunds, although he moved to Norwich with his parents in 1762 at the age of ten.[7] His father was a collector of excise, his mother the daughter of a minor Suffolk squire. He was sent to Holland at the age of twelve in order to learn Dutch, presumably to prepare him for a career in the local textile trade, for on his return two years later he was apprenticed to a textile merchant and later set up in business on his own account. On the death of his parents, however, he used his inheritance to buy a small farm at Sustead in north-east Norfolk, where he spent his time farming, reading, sketching and socialising with local landowners, especially William Windham of nearby Felbrigg Hall (whom he accompanied to Ireland as private secretary and helped in his political career). In 1786 Repton moved to Hare Street in Essex and it was here, in 1788, that he resolved to set himself up as a professional landscape gardener. His first paid commission was at Catton in Norfolk. Suffolk commissions followed soon after.[8]

Unlike Brown, Repton did not employ a team of technicians and labourers. Instead he simply gave owners advice on how their properties might be 'improved', presented in the form of attractive volumes which are normally referred to as 'Red Books' because they are bound in red Moroccan leather. These contain a hand-written text accompanied by water-colours. Some of these have a slide or overlay which, when flat, shows the present appearance of the property, but when lifted depicts the results of the proposed 'improvements'. The Red Books were a clever marketing device, and some contemporaries saw them for what they were. William Mason, for example, described how Repton

alters places on Paper and makes them so picturesque that fine folks think that all the oaks etc. he draws ... will grow exactly in the shape and fashion in which he delineated them, so they employ him at a great Price; so much the better on both sides, for they might lay out their money worse and he has a numerous family of children to bring up.[9]

Red Books survive for eight Suffolk commissions, but Repton's involvement at several other sites in the county is indicated by entries in his account books, the first few years of which survive in Norfolk Record Office,[10] and by references in his published works. For in marked contrast to Brown, Repton was a prolific writer, who penned a number of successful books during the course of his lifetime: *Sketches and Hints on Landscape Gardening* (1795), *Observations on the Theory and Practice of Landscape Gardening* (1803), *An Enquiry into the Changes in Taste in Landscape Gardening* (1806) and *Fragments on the Theory and Practice of Landscape Gardening* (1816).

Repton's style, as we shall see, was more delicate than that of Brown, more subtle and more considered. Moreover, he was acutely aware of the social nuances and implications of landscape design, and he was particularly adept at visual illusion – at hiding, for example, the limited extent of a park. His style was particularly well suited to the kind of small-to-medium-sized estate with which Suffolk abounded; on these his aim was 'to produce great effects without great exertions; or in other words to take the utmost advantage of all natural beauties, with comparatively little expense'.[11]

Repton became involved in a major dispute with the prophets of the picturesque, Payne Knight and Price, who criticised him for closely following the style of Brown. In reality, Repton's style had more in common with the picturesque principals they advocated than their acrimonious debate would suggest. Indeed, this can be seen even in his earliest Suffolk commission – Shrubland Park, near Coddenham, in 1788 – which was perhaps only the second or third paid engagement of his career.[12]

A few months before Repton came here, the estate had been bought from the Bacon family by Sir William Middleton (after 1822 Sir William Fowle Middleton) of Crowfield. Quite how he came to employ this unknown designer is unclear, but by 1789 Repton had produced one of his first Red Books (actually a misnomer in this particular case, since the proposals are bound in a flimsy green paper cover) (Plates 10, 11). The volume typically begins with complimentary comments about the appearance of the park, which it describes as 'beautifully varied by inequality of ground, and richly ornamented with timber of prodigious size'. Repton was, however, more critical of the site of the Hall. This had been rebuilt on a new site some ten years earlier to designs by James Paine, 'near the edge of a steep declivity' overlooking the Gipping valley.[13] This site offered an extensive view, but one which lacked a defined foreground, something 'which is essential in a good composition'. Repton, clearly, was thinking here along broadly 'picturesque' lines: the landscape should be organised like a picture, following the rules of painterly composition, around the 'three distances'. His proposals for the immediate surroundings of the hall involved the creation of an extensive pleasure ground – a 'dressed ground' of grass, gravel paths and shrubs, and he urged particularly careful planting where the ground sloped away steeply to the west. Here there should be 'shrubs towards the summit, and high growing trees at the base', in order to give a feeling of 'safety to the esplanade before the door and make the house appear to stand firmly and proudly embossom'd in wood instead of tottering on the brink of a precipice'. This was typical of the sensible, homely approach Repton always brought to his landscapes, and which contrasted sharply with much of the rather impractical and doctrinaire advice of his picturesque opponents.

Repton also suggested that the woods growing along the escarpment to the north of the hall should be augmented in various ways and that the prospect tower here should have pinnacles added to its apex in order to appear 'more light and pleasing' when viewed from the house. Touches of the picturesque here, perhaps. More striking was the way in which Repton opposed Middleton's proposals to demolish the remains of the Old Hall, which lay some 500 metres to the north of the new house. He argued that part of this structure should be retained as an interesting feature in the landscape of the park: 'In taking down the old mansion, I cou'd wish for the reverence due to antiquity, that some part might be preserved, particularly the chapel and the chequered porch.' The walks which were to run through the proposed new pleasure ground beside the hall could be extended to 'seats in, or near these remnants of ancient Piety and hospitality', while the walled kitchen garden adjacent to the old house should be extended and planted up with shrubberies in order to form a small detached area of ornamental ground.

Other proposals in the Shrubland Red Book are less picturesque in character, but reveal that already, at the very start of his professional career, many of Repton's main ideas had taken shape. He paid particular attention to entrances and approaches – he was supremely aware of the old adage that 'first impressions last longest'. The existing drive to Shrubland Hall led off the Norwich road, curving southwards on entering the park. This line was to be maintained, although Repton urged that alterations should be made to its gradient: it would 'ascend the hill without difficulty when part of the ridge is thrown down' and then run through the new 'dressed ground' to terminate at the house. In addition, he suggested creating an entirely new drive which would approach the Hall from the south, i.e. from the direction of Ipswich. This would run in a grand sweep through the landscape, providing a more striking approach to the hall. It would leave the turnpike road in a gentle curve (a typical Repton touch) at a new lodge flanked by plantations.

The Shrubland Red Book contains a number of other interesting proposals. It recommended various minor alterations to the house in order to improve the views from the rooms within, urged that new woods should be established beside the proposed south drive and elsewhere in the park, and suggested erecting a new gamekeeper's cottage on the eastern edge of the park: 'a delightful object in the approach with the smoke curling in fleecy folds against the dark grove of trees which terminates the landscape'. The latter, again, is typical; Repton was keen to introduce signs of life into the landscape of the park – an approach rather different from that manifested in the monumental exclusivity of Brown's designs. The park was, nevertheless, to be extended, both to the west and to the south, Repton urging that: 'The whole of the grounds through which the approaches pass, be thrown into open lawn, removing the hedges & leaving occasional single trees, or Groups, and foresting the plains with thorns and maples'. Such planting would introduce a measure of busy variety into the scene, more in keeping with picturesque notions than with the sweeping, uninterrupted turf of Brown's parks.

Repton's proposals for Shrubland are of considerable historic interest. What remains unclear is precisely how many were implemented. The Tithe Award maps for the parishes of Coddenham and Barham, surveyed in 1839 and 1840 respectively, and a number of early nineteenth-century drawings and paintings show that much was indeed done as he proposed.[14] The suggested pinnacles were added to the prospect tower, more extensive pleasure grounds were laid out around the house, the planting in the park was augmented more or less in the ways he had suggested, and the new southern drive was put in place. Many more of his proposals, however, seem to have been completely ignored, or else rather loosely interpreted. Thus the gamekeeper's lodge was never erected and the park did not expand southwards until after 1808, or westwards until after 1820.

Interestingly, one of the finest features of the modern landscape seems to have been created through the *partial* implementation of one of Repton's proposals. Although his suggestion of making a path from the house all the way to the site of the Old Hall was not taken up, his description of the first part of its route – 'along the natural Terrace under the spreading branches of the venerable Chestnut Trees' – is so reminiscent of the gravel walk that runs northwards from the house, today called the Brownlow Terrace, that it is hard to believe that this feature was not laid out at this time (it was certainly in place by 1840).

Repton's work at Shrubland was followed rapidly by other Suffolk commissions, his reputation doubtless spreading by word of mouth among the county gentry. He began work at Tendring for Sir William Rowley in October 1790 and produced a Red Book in 1791.[15] Here, again, a park already existed – perhaps laid out when the hall was rebuilt in the previous decade. Repton was rather critical of the house and its location, however, and the approaches likewise left much to be desired: 'The present line of approach from the lodges seems very objectionable, because for a considerable distance it is too nearly parallel with the high road and betrays the narrowness of the park in this direction.' The line of the drive was therefore to be altered, and the other main drive – leading in from the direction of Stoke – was to be improved by the addition of a new lodge at the point where it met the public road. As at Shrubland, Repton thought that the park should be extended and that a larger area of pleasure ground should be created around the house. This was to consist of an area of lawn grazed by sheep and cut off by fencing from the main area of the deer park. He proposed that views should be opened up into the more distant parts of the park by felling trees in the avenue leading up to the house, where at present the mass of foliage served to 'conceal the whole park and lead the eye to distant corn fields which should not form the principal object of a view from the house'. He also suggested that a number of new plantations should be established, 'notwithstanding there is so ample a proportion of wood in Tendring Hall Park'. A wood had already been planted to the north of the house in the previous year – presumably following advice given during his initial visit – in order to provide shelter from the north winds and to hide the house from public view. Two others were now proposed, to 'shut out the village of Stoke'. The tower of the parish church, however, would remain visible, forming 'a very beautiful and interesting feature of the place'. Other clumps and plantations were proposed to obscure views of disagreeable objects or to enhance the existing beauties of the park.

Repton wanted to dam the river Stour, which runs east–west through the park, below the house in order to make a lake. But topography was against him. The width of the valley was such that this could only be achieved by constructing a lengthy dam at considerable expense, while 'it would be equally expensive to dig it out as a river on account of the quick descent; besides, from the lofty situation of the house it would be very difficult to conceal the mechanism of any artificial Water, since a river would require to be of different levels.' An existing formal canal, beside the temple in the west of the park, could, however, be improved by making it more serpentine, in 'a very easy taste', widening it towards the north and giving its western end a fashionable twist.

Lastly, Repton proposed making a number of walks in the area around the house, 'including not only walks of pleasure, but of convenience, as they will lead to those objects which are always interesting about a place'. One should run 'under the lime trees' to the stables and kitchen garden; another might pass through the old fruit garden, across the lawn, terminating at a 'small pool within the wood, which I think may be made a beautiful circumstance in the pleasure ground'. A third might lead to an area which Repton proposed should be used as 'Lady Rowley's flower-garden ... appropriated to all the hardy kinds of botanical plants ...'

Once again, the Red Book for Tendring – although penned very early in Repton's professional career – displays many of what were to become his most typical practices: close attention to drives and entrances, subtle alterations to the planting in order to open up views towards the more agreeable incidents in the landscape and close off those towards the less congenial, and a keen interest in extending and elaborating walks and pleasure grounds.

Livermere for Nathaniel Lee Acton followed hard on the heels of Tendring. The park was first visited in September 1790, and the Red Book finished in January 1791.[16] Typically, Repton was complimentary about the existing scenery with its 'ample lawns, rich woods, and an excellent supply of good coloured water'. In most parts of the park the 'wood and water' were 'beautifully connected with each other'. But there was insufficient vegetation around 'that part of the water where the banks are flat', and also a 'want of clothing' near the house. The latter was a matter of some concern, since 'A large mansion in the centre of a park will never appear with dignity unless it is supported by such a mass of wood contiguous to it, as may hide the disjointed offices and buildings which a large mansion must have in its vicinity.' This was particularly the case at Livermere where, according to Repton, the large wings were rather out of proportion with the main body of the house. The existing screens of vegetation here should therefore be augmented, he suggested, in order to give the impression of a more balanced and harmonious structure.

Additional planting was proposed to the north of the hall in order to obscure glimpses of the arable land beyond the park. The views were to be improved in other ways. The layout of the shrubbery to the west of the house was to be altered so as to open up a prospect from the house towards a particularly attractive group of trees. Repton also proposed alterations to the tower of Great Livermere church, which stood on the edge of the park. He suggested removing the modern lantern, 'a kind of Grecian Cupola' that he considered to be 'totally incongruous with a Gothic tower'. Little Livermere church, which lay within the park, was to have its tower embellished with four pinnacles and its churchyard planted with cedars, cypresses and yews, 'to break the too great length of the building, and act in harmony with the scene' (Figures 35, 36).

More dramatically, changes were proposed to the lakes within the park which (as we have seen: above, p. 58) had been created in the 1750s. The main area of water, the Broad Water, was to be deepened and the dredged earth used to create islands, the intention being to give the lake a less marshy appearance.[17] A number of new plantations were to be added to the shores and, rather bizarrely, the existing boathouse was to be altered so that it would look 'more like a small chapel or chantry'. The Long Water, which extended southwards from the main lake, was to be widened so as 'to render it more conspicuous from the house'.

New walks were to be laid out, particular attention being paid to entrances and approaches. That leading into the park from the east was to be improved by replacing the wooden bridge (presumably that built in 1753; above, p. 58) with a stone one, a new clump was to be planted to the north of the drive and a lodge was to be established at the entrance. The western approach, from the direction of Livermere village, was more to Repton's taste, but here too improvements could be made:

> The other approach … is very interesting: the neatness and cheerful comfort diffus'd through the whole scene is peculiarly pleasing, it expresses affluence attentive to its poor dependents; but the little alterations to the course of the road which I have staked on the ground, would make it more appropriated to the estate; and the same kind of paling should everywhere be used, to mark a unity of property …

This is a particularly interesting passage, as it is one of the earliest examples

FIGURE 35 and 36.
The Red Book for
Livermere, 1791
'Before' and 'after'
views of the east park:
the bridge has been
rebuilt in stone, the
church tower altered,
and the lodge
obscured by planting.

of Repton's practice of what he termed 'appropriation'; that is, the manipulation of the landscape at and beyond the boundaries of the park in order to express the presence and status of the owner. Lodges erected at the main entrances were his most frequent method of doing this, but sometimes – as here – he employed some other recurrent sign or emblem of ownership. As he was later to put it:

> The first essential of greatness in a place, is the appearance of united and uninterrupted property ... There are various ways by which this effect is occasionally produced ... viz. the church, and churchyards, may be decorated in a style that shall in some degree correspond with that of the mansion.[18]

There was already a lodge here, but Repton suggested planting a clump of trees beside the drive a little way into the park; this would both screen the lodge from the house and ensure that the house was not immediately visible on entering the park. As is so often the case, it is not entirely clear how many of Repton's proposals were accepted. Many of Repton's new clumps were certainly established, and one island at least had been created within the lake by the time the Ordnance Survey draft drawings were made in *c.* 1815. A new bridge seems to have been constructed over the Long Water and a lodge erected at the Ampton entrance, but the proposed alterations to the churches and the boathouse never appear to have been carried out.

Repton was busy in Suffolk in 1790. Two months after his first visit to Livermere he was at Henham, producing a Red Book in April of the following year. The estate accounts show that he was paid £48 4s. od. on 8 April.[19] This was, in some ways, a more demanding commission. Henham had been the seat of the Rous family since the mid-sixteenth century but their ancient courtyard house had burnt down in 1773. It was eighteen years before Sir John Rous began to erect a replacement, designed by James Wyatt, on a new site some 100 metres to the south.[20] Repton was called in to create an appropriate landscape for this new mansion; he had already worked with Wyatt at Holkham in Norfolk in the previous year, and was to do so again on a number of occasions. Repton himself seems to have decided the site for the new house, and he may have been instrumental in deciding the rather unusual disposition of its interior. The principal reception rooms were on the first floor: a saloon in the centre, flanked by dining room and drawing room. The choice of a new site for the house and the elevated character of the principal rooms (unusual for the period) both seem to have been motivated by a desire to provide extended prospects in rather flat terrain.

While there was already a park at Henham this contained a number of formal, geometric features, including two avenues. A section of the southern avenue, which was composed of limes, was used by Repton to line a section of the new south drive, but the rest were removed and a more 'modern' park laid out with irregular scatters of trees and a number of clumps. As ever, particular attention was paid to the approaches, and the south entrance was to be supplied with a fine new lodge, flanked by plantations. The south drive, however, passed a number of farm buildings which Repton thought detracted from the status of the place. As it would have been be difficult to screen them effectively (and this would constitute a ruse 'beneath the dignity of Henham Hall') he suggested that they should be redesigned in a more ornamental form 'to give them such an appearance as shall correspond with the general character of the place'.

FIGURE 37. The Red
Book for Henham,
1791 The eastern belt
is broken to allow
distant views of
Southwold.

COURTESY SUFFOLK COUNTY
COUNCIL

Repton was always opposed to the somewhat claustrophobic effect created by continuous peripheral belts, preferring to open them at intervals in order to allow selected views out into the surrounding landscape. At Henham he was keen to cut a wide opening through the 'long unthinned' grove called the Rookery, to the east of the house, in order to create a vista towards the busy, distant port of Southwold (Figure 37). It is not entirely clear how many of Repton's proposals were implemented, but the new drives seem to have been put in place, and some at least of the proposed planting. The park survives, in remarkable condition, although the house itself was demolished in the 1950s.

If many of the key elements of Repton's style were present fully-formed in commissions undertaken for Suffolk landowners at the start of his career, others developed very quickly. Repton began working at Little Glemham in 1791, and in the Red Book introduced, perhaps for the first time, a concept which was to inform much of his subsequent work: that of 'cheerfulness'.[21] Repton hated damp, gloomy residences, and was complimentary about the alterations which had been made to Glemham Hall earlier in the century. These he believed had lessened the 'cumbrous gloom, which our ancestors always annexed to grandeur', creating a building more in the 'modern stile of elegance, convenience, and cheerfulness'. Nevertheless, further improvements were required. The building should be given a stone–coloured wash, he thought, in order to lessen its resemblance to a brick-built 'house of industry' (i.e. a workhouse) and to give its north (entrance) front a less gloomy appearance. He proposed the addition of a central pediment for the same reasons. But the gloomy nature of the place was also, in part, the consequence of its damp and low-lying character, and Repton urged that the square fish ponds in the parkland to the east of the hall should be filled in. Instead a small lake should be created, further away from the house. This would be retained by a dam, which was to carry the minor road that had formed the eastern boundary of the park, and which was now diverted to allow its expansion.

In this case, Repton was against expanding the park to any great extent – expansion to the west would have involved diverting the Ipswich turnpike road, an expensive endeavour. He argued – again, following principals which were to underpin his work throughout his career – that a 'large extent of park in *reality* is not necessary, but it is essential in *appearance*'. The agricultural land beyond the road should therefore be made to appear as a continuation of the park, by replacing the roadside hedges with sunk fences and planting up the fields beyond in a suitable style. The occasional passing carriage would contribute 'to the character of cheerfulness', with the road appearing to run through the middle of the grounds – again, we can see how Repton was opposed to the idea of creating extensive, hermetically sealed, exclusive parkland as the necessary setting for the houses of gentility. And once again, particular attention was paid to entrances, Repton arguing that the line of the existing approach (leading off from the turnpike road to the west of the hall) should be altered and a new lodge erected in order to present an entrance more in keeping with the dignity of the house. To the east of the house Repton recommended removing the stable block ('which looks more like a dwelling house than stable') and thus providing an extended prospect beyond the park, across the neighbouring agricultural land, which would be improved with the addition of a plantation on the neighbouring hill. A new approach, leading in from the east, was also proposed – complete with an entrance lodge 'so constructed as to appear like a tower rising above the wood'.

The Red Books were normally compiled after some initial discussions with the client, and Glemham's owner, Dudley Long North, was clearly keen to retain many of the old-fashioned formal features, including walled gardens and avenues that still existed in the vicinity of the house. Repton thus described how he had 'in compliance with the wish of my employer ... left the strait-mall in the garden undisturbed' and had 'also concurred in another strait walk on a terrace to the south'. The lime avenue leading south from the house beyond the walled garden was also allowed to survive (it exists to this day) but that leading north from the house was removed. We do not know whether North had originally wanted to keep this, but either way it was felled, in accordance with Repton's conviction that this, 'instead of sheltering the house from cold winds, acted rather as a tube to direct and increase their force; besides such tall trees, near the mansion, contributed very much towards the gloom which it is our object to dispel'.

Maps of 1821 and 1826 indicate that many of Repton's proposals were indeed implemented (Plate 12).[22] The fish ponds had been removed, as had the northern avenue, and a new approach from the east had been created more or less along the lines proposed. Above all, while there is no evidence on the ground today of the sunken fences which Repton recommended placing beside the main road to the north and west of the park, the area beyond the road, although still divided into separately named fields (Crabtree Pasture, Jenkinsons, Grays, and Sandpit Hills) had now been laid to pasture and planted up in an aesthetic manner with clumps and scattered trees. Areas of woodland formed a kind of backdrop or outer belt to this area. The line of the western entrance was probably altered in the ways that Repton suggested, and the plantations he had recommended to the east of the hall form what is now the east belt of the park. Although several of his proposals were ignored – the lake was never created and the stables remained in place – Repton seems to have succeeded in his aim of creating a more 'cheerful' landscape (Figure 38). Indeed, Little Glemham remains today one of the finest eighteenth-century parks in Suffolk. It is particularly visible from the turnpike road – now the busy A12 – which still forms its northern and western boundary.

FIGURE 38. Little
Glemham. The thorns
by the south drive are
not specifically
mentioned in the Red
Book, but they are a
classic Repton touch,
and may well have
been suggested by him.

In the early years of his career Repton worked on a number of sites with the architect James Wyatt: Henham is one example, Culford another. Wyatt had been engaged by the Marquis Cornwallis to alter the hall at Culford in 1790; Repton came the following year and the Red Book was completed in 1792.[23] Repton's main concern was to improve the setting of the new house by redesigning the lake. He was critical of the existing area of water, probably created in the 1740s (above, pp. 57–58):

The house stands on the side of a hill gently sloping towards the south; but nearly half of the natural depth of the valley has been destroyed to obtain an expanse of water, which, in so flat a situation, I think ought not to have been attempted: and I am certain,, by proper management of the water, the house would appear to stand on a sufficient eminence above it, and not so low as the present surface of the water seems to indicate; since the eye is always disposed to measure from the surface of neighbouring water, in forming a judgement of the height of any situation.

The lake was to be extended to the west, the water being spread thinly across the meadows rather than dammed to any great depth, so that the current would continue to flow, 'since water apparently in motion as a river is always more interesting than a stagnant pool or lake of small dimensions'. Variety and cheerfulness would always prevail in Repton's landscapes.

His other proposals for Culford followed a pattern which should by now be familiar. Subtle alterations to the pattern of planting were used to close off certain prospects and to open up or enliven others. A nearby church was to be brought into the view: 'There is a church standing on the summit of the opposite hill to which I have ventured to add a spire.' A new kitchen garden was to be constructed a further to the east of its existing site. The most interesting of Repton's proposals, however, relate to the entrance to the grounds and to the proximity of Culford village. He noted that 'village hovels often obtrude on the dignity of the Mansion, and in many cases have been sacrificed without mercy to the necessary parade of solitary pomp.' In consequence, the labourers were 'frequently obliged to walk many miles after a hard day's labour to reach their miserable homes'. Nevertheless, the village of Culford clearly clustered too close to the entrance for polite taste, and in order to remedy this defect Repton suggested that 'the approach from Bury may quit the present high road at such a distance from the village as totally to exclude all symptom of its proximity by planting as a thick cover the whole of the field thro' which the road passes in a narrow line to the Park Gates.'

As at the other places so far discussed, Repton's proposals were in part accepted, in part rejected and in part applied in spirit but not in detail. The main drive was not realigned in the manner proposed, but the Marquis clearly agreed that the proximity of village and entrance was a problem. In 1804 the main road was diverted away from both hall and village, but the village street continued to function as the main approach to the hall. Only later, some time before 1817, was an entrance drive created leading off this new road, which approached the hall to the east and thus avoided the main street of the village.[24] Many of Repton's other suggestions, in contrast, seem to have been adopted completely and immediately. The new kitchen gardens were created between 1795 and 1798,[25] while the lake appears to have been under construction by 1795, for in December of that year the *Bury and Norwich Post* reported: 'Monday morning was found dead in his bed at Culford-hall, Mr James Taite, who was employed in superintending the canal now cutting in Marquis

Cornwallis's park, and who was many years in the service of Sir Charles Kent, Bart., as principal gardener, at Fornham, and some time in that capacity with Nathaniel Lee Acton Esq of Livermere'.[26]

Repton, as we have seen, was always anxious that the grounds of a mansion should reflect the status of the owner. At Broke Hall, Nacton – beside the estuary of the river Orwell – he faced a particularly interesting challenge.[27] He worked here in 1792, his client one Phillip Bowes Broke, who had once attended the same school as Repton. Once again he worked in association with James Wyatt, who remodelled the hall in 1791–93. A small park already existed here – laid out some time in the 1770s[28] – but Repton was concerned that the status of the hall, and therefore the 'Dignity due to the long establishment of the Bowes family', was compromised by the fact that it adjoined the more extensive landscape of Orwell Park. It was imperative that Nacton was given the appearance of a 'united and uninterrupted domain', and much of Repton's design involved careful planting which would screen out the parkland, although not the plantations, around Orwell house.

This concern for status was combined with a desire to provide a more 'cheerful' setting. New plantations were thus proposed, carefully positioned to hide any views of the mudflats exposed in the Orwell estuary at low water and yet 'leave the channel beyond at all times open to the view' (Figures 39, 40). The views to the south-east were in need of particular attention, and here the park was to be extended, plantations established and various changes made to the existing planting. To the east, the nearby village was 'cheerful' enough, but counteracted the 'impression of undivided property' for which Repton was striving. Only if the park could be extended in this direction could the village be admitted to the view, as a 'very beautiful appendage'. Once again, Repton proposed a number of walks through the park, leading not only to the 'conveniences … of a country residence', such as the kitchen garden, but also to the 'striking features of picturesque beauty' in the vicinity.

Repton expressed particular pleasure that the dignity of the hall was being enhanced by the work of his 'ingenious friend', Wyatt, and he was enthusiastic about the architect's choice of the new 'Gothic stile' for the repairs and alterations. This meant that it would be possible to retain the existing approach, along the lime avenue to the west. Such an archaic feature would be 'perfectly consistent' with the house's gothic character, although Repton believed it could be improved by the addition of flanking plantations at the western end. These would give the appearance of passing through a wood for half of the avenue's length, and he also proposed expanding the parkland in the area immediately to the north of the avenue's eastern end.

Some of Repton's suggestions, especially those regarding the area to the east of the house, had probably been implemented by 1803.[29] But other ideas, such as the expansion of the proposed park to the north-east, depended in part on the acquisition of several acres of land belonging to the Vernons of Orwell Hall and these do not seem to have been carried out until the 1820s (a footpath closure order of 1827 probably heralded the start of this work).[30] Many of his

other proposals, however, do not seem to have been adopted at all. Broke may simply have disagreed with them. Repton also worked at other sites, such as Glevering and Wherstead, but much less is known about these projects.

Although Price and Payne Knight would have strongly denied any similarities between their ideas, and those of Repton (whose landscapes Knight described as 'designed and executed exactly after Mr Brown's receipt, without any attention to the natural or artificial character of a place'), it is apparent that many aspects of Repton's landscapes echoed the ideas promulgated by them: the interest in distant views and the hostility towards parks hemmed in by belts of trees, the desire for variety and for signs of human life in the landscape, and the affection for old trees and ancient gothic ruins. In part the debate was an ideological one, for Knight and Price seem to have believed in the creation of landscapes designed to reflect the involvement of the landowner in the local community, whereas the paternalistic concern for tenants and labourers was a sentiment that the fashionable landscape park appeared to negate. This was an important issue at a time when revolution was erupting on the other side of the English Channel. Price in particular bemoaned the disappearance of small landowners and the steady growth of larger estates, a trend strikingly expressed in the development of the landscape park: 'Vast possessions give ambitious views, and ambitious views destroy local attachments" as Price put it.[31] Both Knight and Price thus took particular exception to Repton's concept of 'appropriation': 'There is no such enemy to the real improvement of the beauty of grounds as the foolish vanity of making a parade of their extent, and of various marks of the owner's property, under the title of "Appropriation".'[32]

Only one of Repton's Suffolk commissions – Rendlesham, for which no Red Book survives and about which comparatively little is known – dates from the period after the eruption of the 'Picturesque controversy' in 1794. Suffolk is not, therefore, the place to examine his later works (his career continued until 1816). Yet it is important to note that – under the impact of growing rural poverty and continued fears of unrest engendered by the French Revolution – he gradually came to adopt many of the paternalist sentiments of the picturesque writers.[33] Old techniques were, in the texts of the Red Book, given new justifications. Thus the extension of landscaping out into the surrounding countryside and the creation of strong visual links between the country house and the village and parish church were, by the time that his *Observations on the Theory and Practice of Landscape Gardening* appeared in 1803, no longer discussed principally in terms of 'appropriation'. They were now mainly intended to express the landowners' paternalistic involvement in the community. Repton thus described how the cottages of the labouring poor, if they 'can be made a subordinate part of the general scenery, ... will, so far from disgracing it, add to the dignity that wealth can derive from the exercise of benevolence'. Such sentiments are even more forcibly expressed in Repton's last book, *Fragments on the Theory and Practice of Landscape Gardening*, published in 1816. But perhaps more striking was Repton's growing interest in the 'foreground' of his designs. By 1800 he was advocating the creation of formal terraces between house and park, and soon afterwards formal *parterres* were being introduced into his designs, while his pleasure grounds continued to grow in scale and elaboration. These particular notions were, as we shall see, to be widely adopted in the early and middle decades of the nineteenth century.

FIGURES 39 and 40. The Red Book for Broke Hall, Nacton, 1792. Improvements to the view towards the Orwell.

COURTESY LORD AND LADY DE SAUMAREZ

Picturesque parkland

Today we think of Repton as one of the greatest of English garden designers. But contemporary landowners did not necessarily revere his work in quite the same way, and it is noteworthy that by 1808 the Middletons were asking one William Wood to suggest modifications to their grounds at Shrubland Hall. Wood was a painter of landscapes and minatures by profession, patronised by the Middletons, and his proposals take the form of a lengthy text preserved in the Shrubland archives which – because the accompanying sketches, plans and maps have not survived – is often hard to follow.[34] The document is important nevertheless, for it shows someone earnestly attempting to give a Suffolk park the picturesque treatment. Woods proposed extending the park southwards, much in the way that Repton had earlier advised, and suggested altering the line of Repton's southern drive, so that it left the turnpike road rather further to the south. A new entrance here was to be provided with a lodge and ornamented with decorative planting. More interesting are his proposals for a number of walks running through the parkland, ornamented with planting and embellished with seats. These would take in a variety of natural or man-made points of interest, which would themselves be decorated with additional planting. Thus an old chalk pit would 'contribute a beautiful feature ... if enriched by a few more trees'. One walk – apparently the most important – was to run from the Hall to the prospect tower, which was itself to be embellished with some judiciously placed ivy.

Wood wanted to establish a number of new drives leading through and out of the park and to build lodges at the main gates (none seem to have existed at this time). He also wanted to see much new planting in the park, involving in particular the use of numerous small clumps of trees. Some indigenous species – principally oaks and beeches – would be used. But larch, spruce and Scots pine seem to have been Wood's favourite species, together with Wey-mouth pine. He also had a soft spot for the Lombardy poplar (which, he believed, had 'suffer'd from its familiarity; because ill-judging people have seen it improperly plac'd', and because it was widely planted beside the homes of 'the vulgar inhabitants'). Some at least of his suggestions appear to have been implemented by the Middletons. The park was now, at last, extended south-wards, and this area – known today as the Dentlings – is still characterised by more varied planting than the other, older areas of parkland, featuring Scots pines, cedars, limes and beeches (including some copper beeches), as well as oaks.[35] An emphasis on walks and drives running through the estate, an evident enthusiasm for varied and exotic planting in the parkland, and such touches as the planting of ivy on the walls of the prospect tower all mark Woods out clearly as a designer working in the picturesque mode.

Most early nineteenth-century designers, however, while they adopted some picturesque ideas, generally worked in a more 'Reptonian' style. One was Lewis Kennedy (1759–1842), a London nurseryman who, like Repton, pre-pared lavishly illustrated volumes to show his proposed improvements. His paintings were, arguably, better than Repton's, but the accompanying text was often rambling (and his poems excruciating!). Kennedy's grandfather had established, with James Lee, a famous nursery at Hammersmith in the middle of the eighteenth century, and the Kennedy family continued to be involved in the business until the 1890s.

Lewis Kennedy's only known Suffolk commission was at Livermere in 1815, and it is noteworthy that, as at Shrubland, wealthy landowners were happy to make alterations to Repton landscapes even within his lifetime.[36] It is also noteworthy that, as earlier in its history, the design of Livermere park was considered together with that of neighbouring Ampton; Kennedy informs us that the park 'adjacent with Livermere' had been 'equally consigned to my guidance'. Like Repton, Kennedy normally began his recommendations with praise for the existing landscape: 'In a County which cannot boast of romantic scenery, the Domain now under consideration however may claim much, from a great share of embellished, or beautiful landscape; and the possibilities perhaps of considerably more.' He was particularly impressed by the existing timber in the park, 'now grown venerable, and still of healthy appearance, by its massive grouping, elegant, and finely undulated character'. But improvements were required. The lodge at the eastern entrance, in Great Livermere village, had long existed but Kennedy considered this to be 'altogether unworthy of the ostensibly tasteful, and elegant style, maintained throughout the Exterior and interior decorations' of the mansion itself. He provided a painting showing the lodge as it should be and the plantations which should be established beside it. He suggested that further planting should be made along the edge of the drive in order to improve the character of the approach.

The lake was also found wanting. New plantations were needed around its shore running right down to the water's edge, and some of the existing clumps should be broken to improve the view from the house. The island in the middle of the lake was to be enlarged, 'to assist in forming a diversified scene'. It should be planted with 'quick growing trees', and 'in planting this spot, attention should be paid to place on it such characteristic plants as may lead the feathered race, dispersed over this natural mirror, to a belief, that 'tis to them a safe and secure retreat.' It is interesting to discover that while Repton's proposals for adding pinnacles to the tower of Little Livermere church had not been carried out, they had not been forgotten by the owners: Kennedy concurred 'with Mrs Acton's taste, in preferring the four pinnacles at the angle of the tower, as then, it will stand distinctly characterised from the Church not far distant' (i.e., that of Great Livermere). The church was also to be given battlements, and its door formed into a gothic arch.

Kennedy made many other proposals in his lavishly illustrated volume (Figure 41 and Plate 13). Typically for the period, particular attention was paid to the pleasure grounds around the house. These had in many places grown 'too close, and dense, wanting only the axe to be here allowed its privilege, for admitting the Sun's enlivening ray'. A rose bower should be provided (a common feature of Kennedy's pleasure grounds) and a 'linarium with or

FIGURE 41. Lewis Kennedy, design for Livermere park, 1815.
COURTESY LORD AND LADY DE SAUMAREZ

without pheasantries'. Lastly, he suggested various alterations to the planting in the area between the house and the lake, and proposed building a 'belvedere' as a termination to the view southwards from the house: 'The building purports to represent some rusticated remnant of former days, and its form quite simple, the lower part, or base, serving as a shelter for sheep, from whence a staircase leading to the tower of the belvedere, where the extent of the View is seen.' This, like most of the more expensive suggestions contained in his volume, does not appear to have been taken up by the Actons. But it is likely that many of the more limited suggestions, relating to the shrubberies and the planting in the wider parkland, were adopted, although it is difficult to disentangle Kennedy's work here from Repton's. Certainly, by the early nineteenth century the park seems to have appealed to visitors with a picturesque eye. The lake came in for particular praise, Shoberl for example describing how it 'Winds through a thick-planted wood, with a very bold shore; in some places wide, in others so narrow that the overhanging trees join their branches, and even darken the scene, which has a charming effect.' The parkland around had also clearly been given the 'picturesque' treatment: 'The banks are everywhere uneven; first wild and rough, and covered with bushes and shrubs; then a fine green lawn, with gentle swells, with scattered trees and shrubs, to the banks of the water, and seats, disposed with great judgement.'[37]

The trend towards more varied and complex planting within parks is evident elsewhere in Suffolk. At Thornham, for example, the Tithe Award map shows a number of clumps scattered through the park, and an undated early nineteenth-century notebook gives details of their composition.[38] One, immediately to the north of the church, contained 'Ornamental plants Cedar of Lebanon Lucombe Fulham and Evergreen Oak thorn trees yews hollies junipers purple Beech'. Another, to the north-east of the hall, was planted with birch, common oak, and Lavent oak, with 'the bottom ornamented with Ornamental plants such as Mountain Ash Laburnums purple beech and thorn trees'. Also widely adopted was the idea of having a number of well-defined paths, running through the park and woods, leading to features of interest or picturesque beauty: often utilitarian aspects of the estate landscape, ornamented in an appropriate manner. At Great Finborough, for example, paths led from the hall to the Dairy, south of the kitchen garden: Henry Davy in 1827 described how the building contained 'some fine specimens of old china and the walls are lined with white tiles brought from Holland by Mr Pettiward'.[39]

One of the main reasons why subtle designs, more open to the surrounding countryside and with more varied planting, became popular in this period was probably the steady increase in the number of small and medium-sized parks. This was a widespread phenomenon, but one particularly well-attested in Suffolk. True, some larger parks were created in the county in the late eighteenth or early nineteenth centuries: Haughley, established some time between 1783 and 1811, which covered nearly fifty hectares;[40] Boulge, laid out shortly before 1810, at 55 hectares;[41] and Benhall, probably created around 1810 and covering c. 49 hectares.[42] But most of the new parks which appeared in this period were (at least initially) of modest size, generally between 15 and 30 hectares. This was a period in which, with the rise in agricultural prices and rents resulting from a general population increase, and then the blockade during the Napoleonic Wars, local landowners were doing particularly well. Improving communications allowed grain and other produce to be moved more easily to London, and also encouraged an influx of 'new' London money. William Cobbett, writing in 1830, believed that Suffolk had become a 'highly favoured county', prospering because of its proximity to 'the immortal Wen'.[43] As the local gentry prospered, they accordingly removed any remaining walled gardens, and expanded their diminutive 'paddocks' into small parks. The scale on which this occurred can be easily gauged by comparing the number of parks – and especially small parks – shown on Hodskinson's map of 1783 (Figure 31), with those depicted on the First Edition Ordnance Survey of 1838 (Figure 42).

The small number of large parks created in the period 1790–1840 were generally in the 'traditional' mode, and were in particular extensively belted. Benhall for example was described in 1830 as 'very judiciously placed at a moderate distance from the bustle of the High Road, and seated in a park of some extent, well wooded, and belted by a most luxuriant Plantation for nearly two miles'.[44] The more numerous small and medium-sized parks, however, needed rather different aesthetic treatment: and as they proliferated, the landscape style of Brown and the rest underwent a gradual transformation.

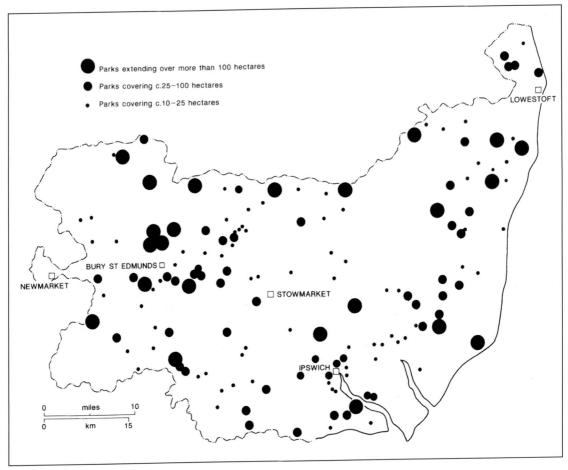

As Repton explained, smaller properties needed more careful consideration than large. It was not by 'adding field to field, or by taking away hedges, or by removing roads to a distance' that the surroundings of such places could be 'improved', but by taking advantage of 'Every circumstance of interest or beauty within our reach, and by hiding such objects as cannot be viewed with pleasure'.[45]

Continuous or near-continuous belts made such parks appear small and claustrophobic; small designed landscapes needed to be designed so as to appear larger than they were, and an increased attention to gardens and pleasure grounds made for much variety in a small space. In all these ways the style of Repton, and of the various picturesque designers, filled a pressing market need, and although by no means restricted to these smaller landscapes was doubtless encouraged by their steady proliferation. The idea of making the designed landscape appear more extensive than it really was, by manipulating the countryside beyond the boundaries of the park, must have been especially appealing to members of the gentry. An interesting, if idiosyncratic, example of this practice can still be seen at Tattingstone, where in 1790 Thomas White built the Tattingstone Wonder – an eyecatcher made from a group of cottages

FIGURE 42. The distribution of parks in Suffolk, *c.*1838, from the First Edition OS 1″ map.

110

to the south of the park. On the south side, which was not visible from Tattingstone Place, the cottages have normal brick walls; on the north side the elevation is of flint, in the form of a church, complete with tower.

One particularly noticeable feature of the period – evident in the advice given by Repton, Wood, and Kennedy presented above – was the proliferation of lodges. The proportion of parks with an entrance lodge, or lodges, in the county more than trebled in the first four decades of the nineteenth century. In part this reflected an increasing concern over security. Rural poverty continued to increase in the early nineteenth century, and the landed classes were perennially concerned about incendiarism and poaching. But lodges also served (as Repton was acutely aware) to advertise the presence of a gentleman's residence, and their popularity perhaps reflects the need for landowners to make their presence felt in this competitive social world.

Suffolk has a number of fine examples of small and medium-sized parks created in the first few decades of the nineteenth century. Glemham House, Great Glemham was built between 1813 and 1823, on rising ground to the north of the river Alde. It replaced an earlier house which lay on a lower and damper site, a little to the south-west.[46] The estate was purchased by one Samuel Kilderbee in the 1780s or 1790s and it was his son, another Samuel, who was responsible for building the new house.[47] It appears that the park – which covered some 30 hectares – was created several years (perhaps much as a decade) before the construction of the new house was begun. A road closure order of 1796 removed a public road to the east of Kilderbee's mansion, probably in preparation for emparking;[48] and the following year Kilderbee received a medal for tree-planting from the Royal Society for the Encouragement of Arts.[49] Indeed, it is possible that the park was originally intended as a setting for the *old* hall. Either way, sales particulars drawn up on the death of Kilderbee in 1829 show that the landscape was typical of its time. The new house lay in the centre of the park, which had belts running around much of its periphery, but with marked gaps to the north-west and in the east, where there were views out across an area of estate land enclosed by an outer belt. It was ornamented both with clumps and with large numbers of free-standing trees, and there was an extensive pleasure ground and flower garden beside the house. The particulars describe how:

> The mansion is approached by a neat entrance lodge and carriage drive, is surrounded by a lawn, with clumps of American Plants, Shrubs and Evergreens, with an invisible iron fence dividing the same from the park.[50]

The park contained an 'ornamental sheet of water' covering about a hectare, complete with a boat house, which lay in the valley to the south of the house. A kitchen garden lay to the north: it contained cross paths which met at an 'ornamental basin', and had a conservatory or vinery 20 metres in length – apparently built against its outside wall, adjoining the pleasure grounds. A path connected the gardens beside the house with an area of woodland, probably a woodland garden, slightly isolated in the park to the east.

The planting in the park is not specified in these particulars, but a rather later sales catalogue (from 1912) mentions oak and sweet chestnut: other species were certainly present, however, for the purchaser, the Earl of Cranbrook, felled 72 Turkey oaks here two years later.[51] Surviving planting from this period, moreover, includes not only oaks (some incorporated from earlier hedgerows, with girths in the range of 4.5–5 metres; other newly planted when the park was laid out, with girths in the range 3.5–4.5 metres) and sweet chestnuts (with girths of 4.4–5.6 metres) but also beech (of similar size) and a number of fine old thorns. The largest areas of woodland, in the north of the park, typically predate its creation: they are ancient, semi-natural woods containing remnants of maple and hazel coppice (now mostly outgrown to form a high canopy) amidst later planting. The belts, however, were planted when the park was laid out, and are mainly composed of oak. The planting beside the main entrance lodge – a small picturesque building with gothic detailing, typical of its time – is elaborate, with box and yew lining the drive, together with a single horse chestnut.

Some original planting also survives in the pleasure grounds around the hall, which were clearly extensive and elaborate: several magnificent beeches with girths ranging from 3.5–4 metres, a scatter of oaks of similar size, a number of fine yews and several examples of holm oak. Much of the holly and hazel growing here is also probably original, and many of the paths threading through the shrubberies survive, complete with their original box edging (although replaced here and there with *lonicera*). Traces of what may have been a woodland garden, isolated in the park to the west of the house, also survive in the form of a box-lined path running through a copse containing lime and horse chestnut amongst much later planting.

Redisham is another example of a Suffolk park newly-created in the early nineteenth century. The estate was bought by John Gardener 'of Westminster' shortly before 1820.[52] The hall, in origin a late sixteenth-century building, did not then have a park but was surrounded by a small area of pleasure ground.[53] Gardener almost immediately demolished the building, and erected the present house, a typical five-bay, two-storey Regency residence with a mansard roof. The park was almost certainly laid out at the same time, and is first shown on the Tithe Award maps for Great Redisham and Ringsfield of 1839 and 1840 respectively.[54] In some ways the park is little different from any created in the previous century: much of the peripheral woodland is ancient, semi-natural coppice-with-standards, and the planting is dominated by oaks, including some relict hedgerow timber but mainly established when the park was laid out, and now with girths of 2.5–4 metres. However, there are a number of typical nineteenth-century parkland ornamentals – especially lime and cedar of Lebanon – and others no doubt once existed.

Glemham and Redisham were medium-sized parks; many of the new landscapes created in this period were much smaller, covering between 15 and 25 hectares. These diminutive landscapes often lacked significant peripheral belts, or clumps; they were little more than areas of pasture with scattered trees, and were not always very clearly distinguished from the surrounding farmland. The majority seem to have been created around existing mansions, although emparking often accompanied their extension or modernisation. Examples include the park at Coney Weston Hall, laid out following a road closure of 1817,[55] and Grundisborough, created between 1783 and 1824, which in 1838 covered little more than 16 hectares, although a number of enclosed pasture fields lay immediately adjacent.[56] In part, as suggested above, the absence or limited extent of belts in such park was a simple function of size: small parks enclosed by woodland would have felt claustrophobic and *looked* small. But it also reflected the new ideas, championed by Repton and others, that glimpses of everyday life (if at a safe remove) could now be considered 'cheerful', and admitted to the prospect. The park at Haughley Plashwood, created some time between 1783 and 1815,[57] covered less than 18 hectares although, typically, a further 6 hectares of timbered pasture lay to the north of the house, on the far side of a public road leading from Elmswell to Wetherden. There were no peripheral belts, and the hall – again, rather typically – lay at the northern edge of the park, thus maximising the apparent extent of the property when viewed from the main, south-facing reception rooms. Davy in 1827 described how the house was

> Situated on a wooded knoll, cheerfully hanging to the south, studded with single trees and groups of ornamental timber. The high road between Bury and Ipswich skirts its southern boundary at an agreeable distance, adding animation to utility and convenience.[58]

Assington, created between 1778 and 1817 and covering *c.* 15 hectares, had a similar open, unbelted appearance. It remained fairly open even after further expansion to *c.* 30 hectares, and the removal further from the house of the kitchen gardens, which had occurred by 1842.[59]

Indeed, some of the diminutive 'parks' created in the decades around 1800 were more in the tradition of the *ferme ornée*, or ornamented farm, than the landscape parks of Brown. That at Holbecks, just outside Hadleigh, was created around 1790, when a road closure order diverted footpaths in the parish of Hadleigh 'so as to make the same more commodious to him the said Sir William Rowley but also to the Publick in general'.[60] Sir William owned the much more extensive Tendring Park a few miles away, and Holbecks appears to have been a subsidiary residence, perhaps used as a dower house. Unusually, the park actually included land owned by and rented from a number of other landowners. Its layout is shown on an estate map of 1811 and on the Hadleigh Tithe Award of 1839.[61] It covered around 34 hectares, lacked clumps or perimeter belts, and was apparently subdivided by fences – it was only distinguished from the adjacent countryside by the fact that it lay under grass (the surrounding land was arable) and was scattered with a number of free-standing trees. It is possible that Repton himself advised on the layout of this landscape. In 1790, the year of the road closure, he was paid for three days work 'at Tendring and Holbecks'.[62]

Not all of the small, new parks created in this period were like these, however. Indeed, the great variety and diversity of designed landscapes in the county is striking. A minority of small parks were much more extensively wooded and belted than these. Sometimes there are obvious reasons for this – as when a residence was in close proximity to a village, so that privacy was a particularly pressing concern. In such circumstances even quite tiny 'parks', such as the 7 hectares Abbey House at Sibton, might be enclosed by prominent belts. But this explanation does not always apply. The park at Marlesford, for example, created between 1795 and 1829 and covering some 20 hectares, had as much as 12 hectares of woodland in substantial blocks and belts around its periphery,[63] while North Cove, probably laid out between 1783 and 1816 and covering around 20 hectares, was prominently belted along its southern side, facing the public road.[64]

As in earlier periods, some of the diversity apparent in landscape design can be related to geographical factors. In particular, a number of new parks were created in this period within or on the edge of Breckland which were both larger than average and more prominently wooded, with extensive belts and clumps. Stowlangtoft park, for example, was laid out some time between 1783 and 1818, and initially covered no more than 16 hectares, but had a thin belt running all around its perimeter and a number of circular clumps, as well as a small lake.[65] The park soon increased in size, for a map of 1824 shows a landscape extending over nearly 60 hectares, containing further clumps and likewise surrounded by belts (although by the time of the Tithe Award in 1843 there had been some reduction in size, to *c.* 35 hectares).[66] Several parks in this area with similar characteristics were established following the enclosure by parliamentary act of what had formerly been completely open, hedgeless country – arable open fields, sheepwalks and heathland. The parish of Brandon was enclosed in 1810, and a substantial allotment some way away from the village was made to Edward Bliss, the lord of the manor. He erected a new hall here between 1820 and 1836, and from 1816 began to plant on a vast scale. Scots pine and European larch, mixed in about equal proportions, were used for the principal plantations, but smaller quantities of *Abies grandis* and Norway spruce were also planted, especially on the eastern side of the property. Here, again, a large number of prominent clumps were established in the open parkland. They were composed of beech, or Scots pine and larch, or a mixture of all three.[67] The park at Exning was likewise established following the enclosure of the village open fields in 1812. The soils here are more calcareous and the planting less dominated by conifers, but nevertheless more varied than was usual in parks of the previous century: a description of 1881 refers to a park 'of about sixty acres studded with remarkably fine Walnut, Elm, Chestnut and other Timber trees, and Clumps of Forest trees of great beauty, encircled by Ornamental Plantations'[68] – featuring, to judge from the trees growing here today, limes and beech.

As in earlier periods, the relatively large size of many Breckland parks probably reflects the low price of land here, while a plethora of belts and clumps was probably considered especially desirable in this still very bleak and open landscape. Plantations provided, in particular, the necessary shelter for game in this inhospitable region.

This was clearly a period in which many new parks were created. But it was also one in which existing parks often increased in size. Polstead Park, for example, which was probably created in the 1760s or 1770s, was expanded to the south-east between 1783 and 1817, significantly bringing the church within the park, a suitable ornamentation for a picturesque prospect.[69] At Drinkstone, the park was extended to the north following the removal of a footpath in 1818, and further expansion took place to the south-east between 1818 and 1839.[70] Ousden Park covered around 22 hectares in 1783; by 1812 it had been expanded to nearly 40 hectares.[71] At Elveden, the park grew considerably in size following its acquisition by William Newton in 1813: over a kilometre of road was closed in the area to the north of the house in 1816.[72] Formerly the hall had stood close to the public road. These diversions, along with the addition of new parkland, ensured a more private position. A similar development can be traced elsewhere. At Orwell, 6 hectares were added to the park in 1815, and a further 10 between 1818 and 1826; these additions, together with two rather complex road closures, again ensured that the house no longer stood close to the public road, but in a more secluded location.[73] Occasionally, expansion followed hard on the heels of a parliamentary enclosure act, as at Worlingham, where the enclosure of Worlingham common in 1787 allowed the small (18-acre) park to be considerably expanded by its owner, Robert Sparrow – probably around 1800, when Worlingham Hall was rebuilt. Much of the former common was planted with trees, creating a large wood. Davy in 1827 noted:

> grounds well-wooded ... The whole greatly improved by the late proprietor Robert Sparrow (d. 1822), who showed much taste in the arrangement of the plantations and the general disposition of the grounds.[74]

By 1840, when the Tithe Award map was made, the park covered *c.* 42 hectares, with a further 50 hectares of woodland extending down on to the neighbouring marshes.[75] The park was clearly designed with picturesque principles in mind. Sales particulars of 1849 describe the gamekeeper's cottage, 'a building of the Norman style of architecture ... designed to serve as a picturesque object from the park', and the 'secluded walks, where the dark and shaggy pine, the sturdy oak and the dense yew are mingled with larch and birch and ... spanish chestnuts', while the ice house was 'screened by dark pines'.[76]

Whether taken into completely new parks, or added to long-established ones, a great deal of land was thus incorporated into parks in the late eighteenth and early nineteenth centuries. Yet there is even less evidence for the removal of settlements by emparking than in earlier periods. Boulge Park provides one of the few examples. The church stands isolated within the parkland, but, as

Hodskinson's map indicates, this arrangement predated the inception of the park. Nevertheless, a map of 1813 shows that a little to the north of the church 'Gilberts Farm House Yards and Orchard' had been cleared away when the park was created a few years before.[77]

Even where parks were not increased in size, many changes were made to their planting and layout, in line with the prevailing fashions. At Hintlesham, for example, Davy described in 1827 how 'great improvements' had been made 'of late', noting the 'embellishment of the house and park by interior decoration and landscape gardening'. The improvements included a very minor change to the public road near the entrance, associated with the erection of a lodge flanked by ornamental clumps, all apparently designed to increase the impact of the approach to the house; and the establishment of a number of clumps and plantations in the fields outside the park to the south-west and separated from it by a public road. Like those planted by Repton at Glemham, these were clearly intended to extend the apparent boundaries of the park.[78] It was in this period, too, that the last of the old 'compartmentalised' deer parks, like Benacre or Loudham, were refashioned in a more modern style. At the former site the subdivisions shown on the map of 1778 (and on Hodskinson's county map of 1783) had been removed by 1840, when the Tithe Award map shows a landscape park of normal form, with a number of prominent clumps and long, sweeping entrance drives.[79] Davy in 1827 described how the 'present proprietor has done much to improve the property'.[80] The deer park at Thornham may also have retained much of its old-fashioned appearance, including an axial entrance avenue, into the early nineteenth century, when much new planting took place (above, p. 109) and road closures permitted a significant expansion to the south. Deer were, however, still kept in many parks, including various foreign breeds as well as the traditional fallow deer. At Helmingham in 1803 there was a 'foreign deer enclosure' in the north of the park.[81]

Gardens and pleasure grounds in the early nineteenth century

As already noted, the planting of gardens and pleasure grounds became steadily more elaborate and varied during the first decades of the nineteenth century, landowners taking advantage of the various new plants becoming available through foreign exploration. In 1807 Sir Thomas Cullum of Hawstead Hall was receiving instructions on how to prepare a border in a garden for American plants.[82] Contemporary descriptions frequently refer to the extensive use of exotics and evergreens. Davy in 1827 typically described the pleasure grounds at Finborough Hall as 'luxuriantly ornamented with evergreens and laid out in good taste'.[83]

Botany and plant collecting, as we have seen, had long been popular hobbies for gentlemen and their wives: indeed, a botanic garden was set up in Ipswich in the early years of the eighteenth century by Dr William Beeston, and taken over by William Coyte, his nephew, in 1732.[84] There is little doubt, however, that this interest grew steadily during the decades either side of 1800 (Coyte published a catalogue of the plants grown in his collection in 1796). One manifestation of this trend was the establishment of the Botanic Gardens at Bury St Edmunds. The collection, begun by Nathaniel Hodson following his retirement from the War Office in 1818, was first located at the eastern end of the churchyard, but in 1831 it was moved to the more spacious site of the Great Court of the former Abbey, donated by the Marquis of Bristol. The central feature of the gardens was an arrangement of concentric, radiating beds, flanked by paths and planted with (mainly native) flowers organised according to their order, species and genus. Although the layout of beds has been much altered, this basic pattern of paths still exists. The collection was maintained by a body of subscribers drawn from among the local gentry and aristocracy, who paid two guineas *per annum*. From the 1840s the gardens became less rigidly scientific and more ornamental in character. A number of trees planted at this time survive, including examples of Fern Leaved Beech, Tree of Heaven, Turkish Hazel and False Acacia.

Even quite small estates might have elaborate gardens, and many owners were enthusiastic gardeners. At Hintlesham, according to Davy, 'the present possessors, Miss Lloyd and Miss H. Lloyd', had done much to embellish the gardens: 'A love of botany and cultivation of flowers has ornamented the gardens with many rare and beautiful plants'. The entrance to the house was by a 63-foot-long colonnade enclosed in glass and forming a conservatory, filled with a choice collection of exotics'.[85]

Indeed, women appear to have taken a particularly active role in gardening in this period, perhaps greater than ever before. When the Mannock family leased Giffords Hall, Stoke-by-Nayland, in 1835 one correspondent praised the new tenants, noting, 'The Colonel ... is very fond of shooting and fishing. The garden, I expect, will be kept in excellent order as the ladies are fond of gardening.'[86]

Sidney Colvin, looking back to the first half of the nineteenth century, described how his mother 'lavished care and money on the beautifying of the grounds and gardens.'[87] Sir Charles Bunbury similarly recalled how his mother 'took great delight in cultivating all the exotic plants she could in a small greenhouse' at Mildenhall in the 1820s, and had an Alpine border there in which she reared a variety of rare species. In 1824 the family moved to Barton where she found: 'A garden ready walled, sufficiently large both for flowers and kitchen garden, and well adapted for these purposes; so she soon formed a very pretty and well-arranged flower garden, in which she cultivated many beautiful and interesting things.' As in earlier periods some degree of gender specialisation is apparent, with women dealing with flowers and shrubs, and men with trees and parkland. Bunbury's father laid out a new pleasure ground 'immediately about the house, separating it from the open pasture which before reached quite to the windows', and planted an elaborate arboretum, formed out of an adjoining paddock and an old bowling green. Between 1823 and 1826 he planted 35 different varieties of tree, 'but it was after his return

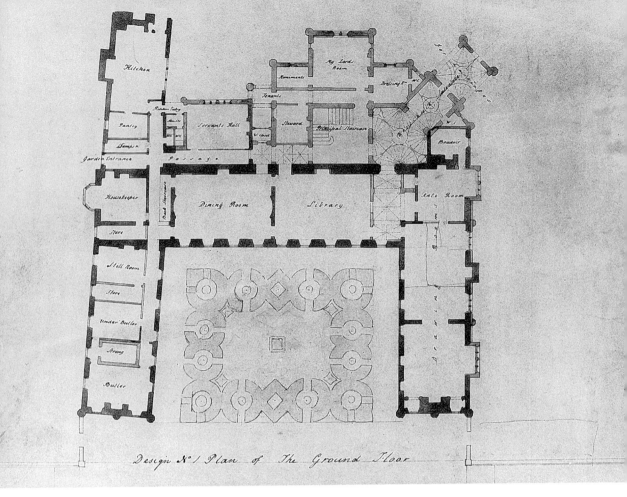

Design No 1 Plan of The Ground Floor

from the continent, in 1830 and the subsequent years, that he did most in the increase of this collection'.[88]

Of particular significance in this period was the steady expansion of shrubberies. Some reached a vast size. At Ampton by 1818 there was one 'of about 20 acres, cut out of ... the park, which is laid out in very just taste'.[89] At Ickworth, similarly, extensive shrubberies were planted to accompany the new hall, which was completed by the Fifth Earl following the death of Frederick Augustus Hervey in 1803. The estate accounts record planting here in 1803 and much work in 1823, including watering shrubs and 'making a new walk'.[90] To judge from later accounts, Roman cypress (*Cupressus sempervirens*) was extensively used, *The Gardener's Chronicle* noting that there were 'literally hundreds of them in all parts of the grounds'.[91] The pleasure grounds here are still impressive. There were further additions in 1847, when a wide range of trees and shrubs was purchased, presumably for the pleasure grounds, including two evergreen oak, four red-leaved beech, six yews, twelve Scots firs, two laburnums, four mountain ash, two arbutus and four variegated hollies.[92]

Following the lead set by Repton in his later career, flower *parterres*, prominently positioned in front of the main windows of the house, steadily increased in popularity. One is shown on a plan of Thornham Hall made in 1837 (Figure 43). They are mentioned with some frequency from the late 1830s

FIGURE 43.
Thornham Hall: the *parterres* in *c.* 1837.
COURTESY LORD AND LADY HENNIKER

118

in sales particulars and other documents. At Fornham in 1843 particulars describe the 'Beautiful flower garden and lawn of about three acres, tastefully disposed and laid out with *parterres* of flowering and other choice shrubs and inclosed from the surrounding park and plantations by an invisible wrought iron fence.' [93]

It was not only small parks which increased in numbers in the early years of the nineteenth century. Even smaller designed landscapes, laid out with care and sophistication along broadly 'Reptonian' lines, made their appearance. These were mostly created by wealthy businessmen or professionals who did not aspire to a true country estate but rather to a small rural property on the edge of town, close to their business interests. Repton himself commented on this development in his last book, published in 1816, noting how it

> seldom falls to the lot of the improver to be called upon for his opinion on places of great extent ... while in the neighbourhood of every city or manufacturing town, new places as villas are daily springing up and these, with a few acres, require all the conveniences, comforts, and appendages, of larger and more sumptuous, if not more expensive places.[94]

The ideas of Repton, Kennedy and the like – the emphasis on detail, on careful and considered planting, and on gardens and pleasure grounds – were peculiarly well suited to such small properties, although in Suffolk at least one notion was widely rejected: these landscapes were generally more closely belted than Repton would have advocated. The grounds of Nowton Cottage, just outside Bury St Edmunds, is a particularly well-documented example. Orbell Oakes, son of the successful banker (and diarist) James Oakes, built up a small estate of 125 hectares on the outskirts of the town between 1801 and 1837 through a long and complex series of piecemeal purchases.[95] His acquisitions included Nowton Cottage, a small farmhouse on the east side of the road from Bury to Hawstead, in 1802.[96] This was soon given a picturesque makeover – illustrations of 1819 show it with rustic porch, thatched roof and dormer windows – and an elaborate, if diminutive, ornamental landscape was laid out around it.[97] To the south of the house was a densely planted pleasure ground; another smaller area lay across the road, with serpentine paths, conifers and area of water and rustic summer houses.[98] By 1827 the grounds had expanded. To the north of the house there was now an area called, presumably because of its shape, the Harp Paddock. This extended across just over two hectares, was belted to north and west, and contained three small clumps. A serpentine path threaded through its western belt. On the far side of the public road was another 'paddock', again belted, with a single central clump, covering some eight hectares.[99] Over the following years the landscape continued to develop. Both paddocks were expanded and further planting made; by 1832 the paddock on the western side of the road covered over sixteen hectares, and by 1840 it was being described, significantly, as 'park or paddock'.[100] Following the death of Orbell Oakes the house was extended by his son, H. J. Oakes, between 1837 and 1840, and renamed 'Nowton Court'. The gardens developed accordingly: an undated plan (probably of *c.* 1837) shows a series of geometric flower beds to the west and south of the house, and to the east an elaborate *parterre*, beyond which lay an extensive shrubbery.[101]

Nowton is a particularly well-documented site, and a fairly typical one; in the vicinity of Bury St Edmunds, Ipswich and, to an extent, Newmarket, similar prosperous 'villas' were springing up in some numbers. The 'singularly elegant freehold mansion' of Thomas Panton just outside Newmarket was up for sale following his death in 1809. The pleasure grounds covered around 12 hectares; they had been 'moulded by the hand of nature into beautiful swelling lawns', and commanded extensive views over Newmarket and the neighbouring downs. The 'rich garden landscape' was 'interspersed with temples, and beset with timber forest trees and thick plantations'.[102] Such developments were not restricted to Suffolk. Indeed, as Repton implies, they were more marked in the vicinity of large manufacturing towns and in the environs of London. Writers on garden design followed the market. John Claudius Loudon began his writing career with a volume entitled *Treatise on Forming, Improving and Managing Country Residences* (1806); in 1838 he published *The Suburban Gardener or Villa Companion.*

CHAPTER 6

Victorian gardens

The return of formal, structured gardens to the main façades of the house culminated in the 1840s and 1850s in the vogue for elaborate architectural gardens in Italianate style – featuring extensive terraces, balustrades, urns and elaborate *parterres* – of the kind associated, in particular, with the designers Sir Charles Barry and William Andrews Nesfield. Conservatories, display greenhouses and arboreta or tree collections likewise grew in importance. This was an age of horticultural enthusiasm, of an expanding gardening press and of gardeners like Joseph Paxton who became household names.[1]

Of crucial importance in all this was the continuing introduction of new species from Asia and the Americas. Commercial nurseries in London and elsewhere bred and disseminated the new introductions on an ever-larger scale, aided by the radical improvements in transport infrastructure, which culminated in a national rail network. But some landowners sponsored their own plant-hunters. William Middleton of Shrubland Hall received specimens sent from Brazil:

> I am most vigorously at work collecting orchids &c., and have got some very good ones which I shall send to Sir William Middleton's in spring if all be well. Bulbs too I have some few, but I should like to have a proper Wardian case sent from England on purpose for them – The attempts made here to imitate these Boxes are wretched failures, and do not answer, and it is a pity that really good plants should run any risk. Ferns too I have got some, but they are difficult things to move.[2]

The enthusiasm for botany and the large-scale introduction of new plants were paralleled by the increasing popularity of formal patterns of planting, and from the 1830s it became the fashion for flowers to be 'bedded out'; that is, grown under glass for part of the year and then transplanted into colourful garden displays. In the 1830s, gardeners employed a wide range of plants (many from Mexico and California) for their floral extravaganzas. By the 1840s, however, a more limited range had become the standard palette: pelargoniums (called geraniums, somewhat confusingly, before the 1860s); lobelias, petunias, verbenas, calceolarias, and salvia patens. Through hybridisation and careful selection the colour range of these flowers was extended to covering six main groups by 1850: yellow, purple, scarlet, blue, pink and white.[3]

There was much discussion in the expanding gardening press about how colours should be combined in beds and *parterres*. When the 'bedding out' system was first developed in the 1830s, particularly under the influence of John Caie (head gardener to the Duke of Bedford), the main organising principal was that colours should be 'clean, simple and intelligible'; that is, presented in solid, contrasting masses, not mixed together. Caie also believed that the size of beds should be considered when the colour of the bedding was chosen. Neighbouring beds of equal size required colours of equal brightness, while a small bed of bright colour could balance a large one of more subdued hue. The height of plants should, he added, be proportionate to the size of the beds. There were many other rules.[4]

Suffolk has a special place in the history of bedding-out plants and of the theories about how they should be combined in colourful displays. One of the key figures in these developments was the gardener Donald Beaton, who came to Shrubland in *c.* 1838 and worked here closely with his employer, Lady Middleton, until 1849. Together, and in parallel with similar work going on at Trentham in Staffordshire, they pioneered two important innovations: the ribbon border (a long narrow bed arranged in rows) and the practice of 'shading'. This was a revolutionary way of organising colour in a bed, invented (according to Beaton) by Lady Middleton herself. Rows or groups of plants of a similar hue were planted next to each other, to 'blend so perfectly that you cannot tell where one ends or the other begins', thus creating seamless spreads of colour.[5] Beaton retired from Shrubland in 1850 and, as editor of the *Cottage Gardener*, and adviser on bedding at Kew, became Britain's principal spokesman and theorist on bedding and colour.

The gardens at Shrubland

The gardens at Shrubland were, unquestionably, the most elaborate and famous in the county. They were the creation of a string of able gardeners and designers working in close collaboration with the owners, Sir William Fowl Fowl Middleton and his wife, who succeeded to the estate in 1829. Between 1830 and 1832 Shrubland Hall was extensively remodelled by the architect J. P. Gandy-Deering, who provided Paine's house with a more imposing entrance and altered the fenestration in order to create a more fashionable vaguely 'Halianate' façade, with prominent balustrades. More extensive alterations were, however, undertaken from the late 1840s, when Charles Barry extended the building and gave it a more rigorously Italian feel. Barry augmented Gandy-Deering's central feature, an open terrace at first floor level linked by stairs to the garden below, adding imposing balustrades and the striking asymmetrically-placed tower which still dominates the building and, indeed, the landscape of the park.[6]

Elaborate gardens were gradually developed through the 1830s and 1840s, largely under the direction of Donald Beaton. The house, as we have seen, occupies a prominent position above a dramatic escarpment. The new gardens,

FIGURE 44.
Shrubland Hall: the
Swiss Cottage.

created in the 1830s and 1840s, were mainly laid out at the foot of this slope with only limited areas of balustrade garden beside the hall itself. They included a maze of box designed by Beaton himself, an elaborate rosery, a Swiss cottage (Figure 44) and alpine garden, and the elaborate Fountain Garden – one of the places where the practice of shading was developed by Beaton and Lady Middleton. This was composed of a large number of radiating beds, each 'ray' being divided from its neighbour by a strip of grass, and sub-divided by concentric grass strips into three separate beds. The beds were shaded with the brightest tint near the centre, the darkest towards the circumference. The plants declined in height from the outer circumference towards the centre so that the planted surface appeared bowl-shaped.[7]

The Fountain Garden was, and still is, bounded to the west by a long curving wall, variously described as the Hot Wall, the Conservative Wall, or the Conservatory Wall. It is 60 metres in length, and rises to a height of over three metres on the eastern side and nearly five metres on the western. It is hollow, and was heated with hot water pipes. The west side was covered with glass, which was removable and taken down in summer. The borders on this side were also covered with glass. Under Beaton's management this side of the wall was planted with Tacsonias 'and other climbers', and the border with African bulbs. The eastern side of the wall was not covered with glass, and here 'plants of a more hardy nature, including Roses etc.' were grown.[8]

To the west of the hot wall were further areas of garden, which have not survived, including the Poplar Garden. This was an area of raised beds where Beaton experimented with 'shot silk bedding', a particularly complex and intricate arrangement of plants of different hues and sizes.[9] All these various gardens were connected by straight and winding paths, running through lawns ornamented with shrubs and urns. They were quite separate from the gardens around the hall, from which they were reached by paths running down the steep escarpment, which was by this time thickly planted with yew, box and a variety of evergreen and deciduous trees

The gardens were thus already elaborate and famous by the 1840s; indeed, the annual bill for their maintenance in 1847–48 was in excess of £1,195 – a huge sum.[10] But they became more elaborate still with the arrival at Shrubland of the architect Charles Barry. Although he was originally commissioned to alter the hall, he already had considerable experience of garden design, having created elaborate terraced gardens in Italianate style at Trentham in Staffordshire and Harewood in Yorkshire.[11] Working with the Middletons, he pulled the various existing garden features into a coherent whole, added a number of new elements and then connected the whole ensemble more firmly to the house itself.

First, from 1850, the string of gardens at the foot of the escarpment was linked together by the Green Terrace, a long, ruler-straight grassy walk (Figure 45). This was widely admired by nineteenth-century visitors, most of whom emphasised that it was nearly a mile long, although 'about one-third

FIGURE 45.
Shrubland Hall: the Green Terrace.

of the length is beyond the garden boundary on the north side'; that is, it extended out into the park.[12] The terrace was bordered on its east (at the base of the slope) by a fuchsia hedge for the first few metres and subsequently by 'a sloping bank of different coloured Dahlias, five or six rows deep, each whole row being of one colour. At other points … not so lighted up with flowers, the shrubs sweep down'. The western side was more formal in appearance, 'margined by parallel beds of Salvin, Irish Yews, Arbor-Vitae, vases raised from the ground, filled with choice Geraniums &c.'.[13]

Next, a number of new areas of garden were laid out immediately adjacent to the terrace. These included, towards its northern end, the French Garden. This was a box-work *parterre* on gravel, featuring a number of beds of roughly similar size: 'the finer lines and scroll work, instead of being planted, are filled between the box with silver sand'. It was surrounded by a thick laurel hedge ornamented with marble busts in recessed niches.[14] At the far, southern end of the terrace the Box Terrace was created, a long thin *parterre*, described by Beaton in 1856 as 'the finest Box terrace in England'.[15] An article in the *Florist* described it as 'completely in the parterre style of the French', and noted that the interior was planted with 'very dwarf flowers': 'Silene Schafi, dwarf French Marigold, Lobelia ramosa, &c.'.[16]

The most impressive addition of the 1850s, however, was the magnificent flight of stairs called the Descent (Plate 14, 15), which ran – and still runs – between two great terraced and balustrade gardens: the 'Balcony Garden', which replaced the existing garden beside the hall, and the Lower or Panel garden at the foot of the escarpment, on the same level as (and crossed by) the Green Terrace. It is usually assumed that this magnificent composition was entirely the work of Charles Barry. But in a letter of 31 August 1850 the architect reported to Sir William: 'I approve of Lady Middleton's plan for the Lower Grand Terrace [i.e. the Balcony Garden] and the proposed termination of it with the seat', and on 31 December the following year he explained that he had

> no objection to your present proposal of a Villa d'Este descent from the Pavilion opposite to the centre of House to the Gardens below; on the contrary I believe that many advantages and great effect would result from it: but it will I think entail upon you the necessity of making considerable alteration to the laying out of the Lower garden in order to create a feature of sufficient importance and effect to be worthy of such a descent.[17]

Indeed, as the wording of the letter implies, the design may itself have evolved gradually. Originally, it seems, the new terraced gardens around the hall were to be linked to the gardens at the foot of the slope by a less impressive flight of stairs located further to the south, opposite the Fountain Garden.[18] Certainly, surviving correspondence makes it clear that construction of the Balcony Garden was begun before the final form of the connecting stairs had been decided. Either way, the contract for the Balcony Garden was awarded to Lucas Brothers, stonemasons of Vauxhall (London), in December 1850.[19] The firm had already been responsible for executing most of Barry's alterations to the house. The estate itself prepared the foundations and the basic substructure.

When completed in 1853 the new garden was a magnificent sight. Beaton described how 'The Balcony Garden is now, as it were, part of the house; being joined to it and the inner terraces, at either end, by a rich system of balustraded stonework, in Sir Charles Barry's most florid style.'[20] The pattern of planting in, and possibly the physical structure of, the principal beds seems to have evolved during the course of the 1850s. By the middle of the decade the main features were a central walk, flanked by four large beds on each side.

> These have raised stone borders, wide and massive; next a band of turf; and between the turf and box edging inside a band of white sand. This gives a high architectural finish to the garden, and is, besides, quite in keeping with the walls, balustrading, vases and other accessories which surround it.[21]

The main object, according to one visitor, was to have 'masses of colours; hence it was imperative that the beds should be large, to produce a grand effect, and warm colours are only employed for the like purpose'.[22] The beds were mainly planted with pelargoniums, with smaller numbers of other plants (including love-lies-bleeding and petunia). Four were scarlet, two purple and two blue. The beds were arranged 'with the plants sized so as to appear no higher than in the exact proportion of a gentle slope from the outside to the very centre of the bed'.[23]

Two long, square turf plots flanked the beds on each side, decorated with a tracery pattern in white sand. The garden also contained further, smaller beds, containing yuccas and hydrangeas, together with a number of stone boxes planted with humeas. Rather larger boxes, ranged along each side of the central walk, were planted with Portugal laurel in imitation of orange trees, 'for which they are excellent substitutes'. The borders under the retaining wall at the top nearest the house were planted with hollyhocks in lines, fronted with Lady Middleton geraniums, 'a rosy scarlet raised here by Mr Beaton, and very valuable for bedding purposes'.[24]

On the far side of the garden is a small stone gateway, the Pavilion, which gives access to the top of the Descent. This magnificent feature was begun in 1852. Its construction on such a steep slope posed a number of technical problems and was preceded by much discussion of methods and materials among the estate staff, for once again Lucas brothers were only responsible for the stonework superstructure. In January 1852 William Davidson, Middleton's clerk-of-works, wrote to his employer to point out that 'The inferior bricks in this part of the county are generally soft and half burned, and will not stand the frost. Cubes of chalk are inadmissible for the same reason.' He therefore suggested that the retaining walls should be constructed of 'flint stones and Concrete', estimating that 444 cubic yards would be required, which at 5*s.* 6*d.* per cubic yard would come to a total of £122 2*s.* 0*d.*,

Including the facing of the flint stones but not including the wheeling and ramming of the earth to the back of the wall which should be done as the constructing of the wall proceeds. A framework of scaffold poles, holding in place planks laid edgeways, should be used to retain the concrete until dry.[25]

Davidson had other matters on his mind. The adoption of this dramatic new plan meant that alterations had to be made to some of the gardens created during the previous decade at the foot of the escarpment:

A straight line from the centre of the Pavilion will cut through the NE corner of Rosary and will very little interfere with what we have done there, and which was finished before I was aware of the intended plans. It will cut through the hypericum bank, but this is a plant easily moved.[26]

By 1 February, however, he was warning Sir William that the new Descent would 'come directly in the way of the Moss Summer House', and enquired whether the 'rearrangement of the Garden below will include the removal of this'. (It was removed to a new site, and later in the century moved again; it still survives, known today as the Moroccan Summer House.)[27] Subsequent letters from Davidson include much further discussion about construction methods, and in the summer of 1852 – when work on the Descent was finally under way – he noted, ominously, 'Where the chalk lies very deep I shall fill up with concrete for a certain height so as to economise on the quantity of brickwork as much as possible.'[28]

Construction of the Lower or Panel Garden began after work on the Descent had been completed, early in 1853. The garden formed a magnificent termination for the Descent, and featured lavish ballustrades, central 'loggia' and splendid *parterres*. The details of the latter were altered on a number of occasions, but their basic layout was described by D. T. Fish in 1857:

The main features are these: – A large artistic bed on each side of the fountain; a broad turf avenue, crossing the central one, embellished with long beds of yellow Calceolarias; then a sunk *parterre*, from which this has been called the *panel* gardens, with a broad avenue beyond, next the balustrades, decorated with similar long beds of scarlet and pink Geranium; whilst the ends are flanked with Petunias and Hollyhocks.[29]

By the late summer of 1855 the new gardens had been completed: as well as the various features already described, further garden areas had appeared beside the Green Terrace, and to the south-west of the Poplar Garden a new rosery had been established to replace that destroyed by the construction of the Descent. The finished works were exhibited to the public, as the local newspaper described:

It will be remembered, that the residence of Sir Wm Middleton was honoured by a visit from His Royal Highness Prince Albert, during the meeting of the British Association at Ipswich in 1851, when the grounds were generously opened to the public. Since that time several additions and improvements have been completed, and an opportunity recently afforded by the liberality of the proprietor enabled a large concourse of visitors to inspect and admire them. On the 24 August, 1855, a grand promenade was held at Shrubland, in aid of the East Suffolk Hospital, of which Sir Middleton is President.[30]

But it was, apparently, even later than this that the final additions to the composition were made. Below the Loggia an area of 'wild garden' was laid out. This contains a massive stone 'rustic bridge', 'thrown over a chasm, where a wild luxuriance prevails, as if nature had been partly left to her own dictates'.[31] As an article in *The Gardener's Chronicle* for 1867 put it, 'By one of those rapid transitions for which the late Lady Middleton was famous, undressed nature, in the form of a rough wooded scene, creeps right up to the base of the highly artistic wall.'[32]

As we might expect, the completed grounds at Shrubland were phenomenally expensive to maintain. This was largely on account of the bedding schemes which featured prominently in most areas of the gardens: already elaborate and impressive under Beaton's custodianship, they became yet more extensive under his successor Mr Foggo. The latter informed the magazine *The Florist* in 1856 that 80,000 geraniums, verbenas, petunias, lobelias and other plants were annually required for planting out in the various beds, borders and vases, 'and this independently of annuals, &c., raised from seed, which are likewise worked into the general arrangement'.[33] A series of garden bills survives for the years 1857–59 and 1864–85. These show some variation over time, but the total expenditure generally hovers slightly over £2,000 per annum (the highest total is for 1877–78, when the overall bill was in excess of £2,579).[34]

Sir William Fowl Fowl Middleton died without issue in 1860, and his estates in Shrubland and elsewhere passed to his nephew, Sir George Nathaniel Broke, of Broke Hall, Nacton, who took the name of Broke Middleton. The inheritance came with certain terms, however. Sir George had to agree to maintain 'the house grounds and gardens in the order in which they were left by the late W. F. Middleton'.[35] The Trustees of the will were to pay him the annual sum of £2,000 to cover the necessary costs. The agreement also stipulated that Sir George should continue to use the firm of Lucas Brothers for any repairs required to the stonework of house or grounds.

Unfortunately for Sir George, twelve years after he inherited the estate it became apparent that drastic repairs were needed to the gardens, and in particular to the Descent. On 17 May 1872, Lucas Brothers wrote to Sir George:

> Agreeably to your letter of the 9th April, we have thoroughly examined the stonework of the House and Grounds, Shrubland Park. We enclose a report of the works we consider necessary to be done ... We strongly advise you to adopt the course we recommend in our letter of the 23 August 1860 – and had this been carried out, which we had engaged with

the late Sir William Middleton to do, the present outlay would have been unnecessary.[36]

The attached report details at length the numerous defects in the garden, and Lucas Brothers were understandably keen to shift any blame for these away from themselves:

> The late Sir Charles Barry and ourselves were very anxious to carry out these works in Portland Stone, but the late Sir William Middleton preferred the great saving of cost of using Caen stone ... Portions of the work require to be made good through the subsidence of made ground and the brick foundations:– this portion of the work was executed by the late Sir William Middleton's own workmen – as also the brickwork alluded to in the report.

The letter suggested that on top of the repair bill of £3,590 (which explicitly excluded a list of 'sundry works ... it is presumed will be carried out by the Estate Workmen'), Lucas Bros. would be prepared to maintain the works at an annual cost of £200.

Sir George was understandably unhappy about this estimate and sought alternatives. He received a much lower one from George White of Vauxhall Bridge Road, Pimlico: £1,896 for the repairs in the gardens, plus an extra £637 for the repairs to the house, making in all a more manageable total of £2,533. White was engaged to undertake the work on 9 October 1872, and George Evans was employed as supervising architect. On 3 July 1873 Evans was able to report favourably on progress, and also to place the blame for the problems fairly firmly at the feet of William Davidson and the estate workers twenty years before:

> When cutting away the Terrace Wall immediately in front of the house, I had an opportunity of seeing the character of the *old work* & I never saw aught more disgracefully done ... If there was any Clerk of Works when the Terrace were built he must sadly have neglected his duty.

The Descent and terraces were systematically repaired, and today survive in excellent condition, arguably the finest piece of nineteenth-century garden architecture in all East Anglia.

Nesfield and others

The gardens at Shrubland Hall were, by common agreement, the most magnificent in Victorian Suffolk. But those at Somerleyton Hall ran them close. The great seventeenth-century gardens, described in Chapter 3, had been removed in the eighteenth century and replaced by much less elaborate pleasure grounds. In 1844, however, Sir Morton Peto, railway entrepreneur and senior partner in a major civil engineering firm, purchased the Somerleyton estate for £86,000. He immediately began rebuilding the hall in grand Italianate style, and remodelling the gardens.[37] The main area of gardens lay to the north of the hall and was ornamented with a number of statues, the work of John Thomas, the architect of the house. The focal point was, according to Crowes's *Lowestoft Handbook* of 1853, a fountain 'Composed of dolphins supporting shells, from which springs a conical shell sustaining a marble statue of the 'Water Lady', holding a lily in her right hand, which throws out a jet of water'. A straight walk – flanked by rose trees and flower beds – led through the centre of this area to the kitchen garden, which was entered through an ornamental archway. Beside the hall, to the south of the garden, there was a large conservatory or winter garden, also designed by Thomas.[38] It was still under construction in 1853, when Crowe's *Handbook* described it:

> At the north end of the Hall, Mr Peto is now erecting (from Mr John Thomas's design) a 'Winter Garden', which will be lighted with gas – having a ring of gas jets around the interior of the dome. The length of this garden (chiefly of iron and glass) is 126 long by 107 wide, height to apex of central dome, 55 feet.[39]

These pleasure grounds had barely been completed, however, when they were radically transformed.[40] Some time between 1854 and 1857 most of the statues were removed and relocated elsewhere in the grounds, the fountain was moved into the winter garden, the pattern of paths was made more serpentine and complex, and a number of exotic specimen trees were planted. Some of the trees growing in this area today probably date back to the 1850s, including fine examples of London plane, sweet chestnut and beech, and magnificent examples of cedar of Lebanon and Wellingtonia – the latter a real novelty, for it had only been introduced into England in 1853.

These changes may have been effected under the direction of the famous garden designer William Andrews Nesfield. He was certainly responsible for a number of other alterations and additions made to the grounds of Somerleyton in this period. Nesfield had begun a career as a watercolour painter after retiring from the army, but became involved in garden design and gradually built up a thriving practice through the 1840s.[41] His particular speciality was the creation of elaborate box *parterres*, based (more loosely than he claimed) on seventeenth-century originals. At Somerleyton a particularly fine example was laid out to the west of the house. It was absolutely typical of Nesfield's creations, defined by hedges of box separated by white stone chippings and flower beds: 'The white gravel forms a good contrast to the dark green of the Box, and brings out the figures in bold relief.'[42]

Like most other examples of Nesfield's work, the *parterre* was laid out on a low terrace surrounded by balustrades, which was terminated by a semibcircular apse-like arrangement looking out over the park. Here there was, and still is, a globe dial (another standard Nesfield touch), described in the Sale Particulars of 1861 as 'An EQUATORIAL SUN DIAL, brass, with Zodiac belt, gilt, supported on a Pedestal 8-ft high, of elegant proportions (Marble, finely sculpted)'. The balustrade surrounding the garden was surmounted by twenty-six urns and a variety of sculptures (again by the architect Thomas).

The *parterre* was flanked by raised terraces adorned with Irish yews, junipers, yuccas, rhododendrons and standard laurels; marble vases on pedestals; and stone flower baskets. The hard landscaping – the gravel paths, terraces, balustrading and much of the stonework – survives, but the *parterres* were grassed over early in the twentieth century, as tastes changed and maintenance costs rose (Plate 16). Fortunately, Nesfield's other great contribution to the landscape at Somerleyton still survives intact and can be visited and enjoyed today: the great yew maze, to the north-east of the pleasure grounds (and east of the hall), with a central mound surmounted by a small pagoda. His design for this impressive feature still exists at the Hall (Figure 46).

Sir Charles Barry designed comparatively few gardens, and only one in Suffolk. Nesfield, in contrast, was a prolific garden-maker, turning out large numbers of rather stereotyped designs, almost all featuring elaborate *parterres*

FIGURE 46.
Somerleyton Hall:
design for a maze,
William Andrews
Nesfield, *c.* 1854.
(COURTESY LORD
SOMERLEYTON)

131

set on low, balustrade terraces. The county can boast a number of examples
of his work, including Woolverstone beside the estuary of the River Orwell.
The main feature of this design was, once again, a broad *parterre*, with terraces,
central fountain, balustrades and urns laid out immediately in front of the
north façade of the house. Beyond, enclosed by balustrading and a substantial
ha ha, and featuring two transverse terraces, was an extensive pleasure ground
stretching away towards the shore of the estuary. A sales catalogue from 1937
described:

> Pleasure gardens of 14 acres sloping to the river Orwell, approached by
> stone steps from the house ... enclosed by a dwarf stone balustraded wall
> with wrought iron gates giving access to the park on either side. They
> are most attractively laid out in a series of terraced lawns and include an
> ornamental sunk, box-bordered bed with central pool and stone figure,
> sunk rose gardens, rhododendron and azalea gardens and a particularly
> fine bed of Kalmia latifolia, hydrangeas and flowering shrubs and
> climbers.[43]

The Gardener's Chronicle for 1867 was positively ecstatic about the layout of
the main *parterre*:

> I have never seen a happier combination of Box embroidery, flowering
> plants, gravel and Grass than in the garden at Woolverstone. It is perfect
> of its kind. Mr Nesfield will excuse me for saying that it is the masterpiece
> of all his productions, and I have seen many of them.[44]

The article also gave a glowing account of the conservatory, the 'outside
fernery' (a rockery cliff constructed somewhere within the park), and the
kitchen garden. Unfortunately, only the barest outlines of the hard landscaping
survive, in what is now the grounds of a private school.

Nesfield also supplied designs for Henham Hall. A text accompanying his
preliminary drawings provide interesting insights into his thinking, and indeed
into attitudes to garden design more generally in the middle decades of the
nineteenth century.[45] Nesfield criticised the fashions of the previous century,
when 'decorated ground near a house was entirely ignored, as an unjustifiable
encroachment on nature'. Indeed, he suggested, with some exaggeration, that
this principle had been followed so rigidly that 'a flower garden was considered
a deformity & banished to some hidden and vacant spot in a shrubbery'. He
approved of the 'refined period of Chas. 2nd', when a 'Decided separation
was invariably designed between Art and Nature ... by means of enclosed
spaces properly proportioned, & that such enclosures contained highly defined
Gardens so disposed in detail as to be *directly* commanded from public rooms.'

His discussion then turned to the particular case of Henham. He noted
that there were some enclosed areas to the east and west of the hall, but
thought these 'quite nonaffective as regards the public rooms'. The main
parterres should therefore be laid out to the south of the hall, where the
principal rooms were located. There was, however, a problem. The house plan

had been developed 'during what may be termed, the questionable manner of the last century', so that the principal entrance was also located on this front: 'Thus the only safe measure is to construct dressed ground on the same principal as the house arrangement – therefore, as the entrance is on the same front as the public rooms so may a forecourt by way of harmony be annexed to a Garden.' Either an axial approach could be flanked by areas of garden, or the approach could be from one side, with one area of garden set asymmetrically to the main elevation – this was Nesfield's preferred solution. Nesfield concluded by expressing a view which, by the 1850s, would have been widely shared:

> When the character of the Park scenery immediately in front of the house is taken into account, it is obvious that the present monotonous and bald foreground is very objectionable, whereas a *parterre* of ample dimensions would not only remedy this striking defect in a most cheerful manner (especially in winter without flowers) but would by the contrast of formality improve the distant scenery.

Nesfield's designs for the gardens here seem to have been carried out. Among the estate papers deposited in the Ipswich Record Office are a 'working plan for the Little Parlour west', showing a *parterre* similar to that illustrated on the document discussed above and probably in Nesfield's hand, along with another, described as 'working drawing of garden' signed by one W. B. Thomas.[46] The hall was demolished in the 1950s and little trace of the gardens survives on the ground today.

Nesfield seems to have carried out a number of other commissions in Suffolk, for among his personal papers (preserved in a private collection in Australia) is a series of county maps, on each of which several country houses are boldly ringed in blue pencil.[47] Unfortunately, it is not always clear whether the places in question represent executed commissions, or simply potential customers, but the map for Suffolk certainly features a number of places in addition to those just described. Flixton Hall is almost certainly an executed site, for the earthwork remains around the site of the demolished hall (Figure 57) are very much in Nesfield's style. Glevering and Bradfield Combust are likewise almost certainly real commissions, for both today boast terraced gardens of mid-nineteenth century date. In the latter case, the hall was rebuilt, by A. J. Young, in 1857. At Orwell Park there are few obvious signs of Nesfield's work, but the house was extensively altered by the architect William Burn, with whom he frequently collaborated, in 1851–3. Burn designed an orangery which survives, although in degraded form, so it is quite likely that Nesfield advised on the grounds.[48] Tattingstone, Sproughton Hall, Brampton Court, Helmingham, and the Chantry at Ipswich are also circled on Nesfield's Suffolk map, although the character of his work here, if indeed there was any, remains uncertain.

Hengrave is an intriguing case. According to an article in *The Gardener's Chronicle* in 1881, a flower garden to the west of the hall '... tastefully laid out in the usual geometric style', had been designed by 'the late Mr Nesfield', a claim repeated in a subsequent account, in 1907.[49] This was not the first garden in the area to the west of the house – the Tithe Award map of 1839 appears to show an oval *parterre*, cut with geometric beds.[50] The broad outlines of the garden described in *The Gardener's Chronicle* still survive although the details have been simplified. It has terraces to east, north and west, and cross paths (flanked by yews) which meet at a central pond. These paths are now paved with stone, although late nineteenth-century photographs show that they were originally gravelled. It certainly looks vaguely like a Nesfeld design, but there are grounds for doubting the attribution in *The Gardener's Chronicle*. Hengrave does not figure as one of the sites circled on Nesfield's county map. Moreover, a large watercolour in the Hengrave Collection in the Cambridge University Library shows 'Hengrave Hall. Plan for proposed Geometric Garden'. It carries the date July 1858 and is signed not by Nesfield but by one James Howe. An accompanying drawing, made in February of the following year, is a working plan for the central section of the *parterre*. Another plan, drawn up in November 1860, shows 'Revised plan for proposed arrangement of Flower Garden and Upper Terrace'.[51] It is possible that these represent modifications to a garden recently created by Nesfield – much other work appears to have been going on in the gardens in the 1860s, and the estate accounts for this period show large amounts of money being paid to major nurseries in London and elsewhere (especially those of Lee, Waterer and Veitch).[52] But given the extent of the work indicated on the plans, it is more likely that the *Chronicle*'s attribution is erroneous. After such elaborate *parterres* went out of fashion, there may have been a tendency to attribute any notable example to their most famous proponent.

James Howe is not a well-known garden designer, and Hengrave is a useful reminder that Barry and Nesfield were not, of course, the only garden designers operating in Suffolk in this period. At Thornham, for example (Plate 17), extensive areas of formal bedding and *parterres* were laid out to the south and west of the hall, while to the north – in a pattern rather reminiscent of Somerleyton – an extensive area of pleasure ground and shrubbery occupied the area between the house and the kitchen garden, which is shown in some detail on a plan of *c.* 1845.[53] The *parterres* at Broke Hall, Nacton, were particularly elaborate, featuring both geometric shapes and a more flowing arrangement of beds: the Crown Garden. Plans in the Ipswich Record Office record the development of the planting in the *parterre* beds over a period of ten years from 1844 to 1854. As well as the customary pelargoniums, verbenas, lobelias and salvia patens, a range of bulbs was also used – 'tulip of sorts', hyacinths and crocuses.[54]

The gardens at Orwell Park, as described in the 1870s, were particularly elaborate. As noted above, Nesfield may have had a part in their design, but they do not seem to have featured one of the great terraced *parterres* which were his hallmark. To the south of the house was an area of lawn, separated from the park by a ha ha, which had been lowered in order to improve the views towards the river from the ground floor rooms. Its surface was 'very judiciously broken up by leaving large, irregular, natural-looking mounds as bases for the many fine trees to stand on' – cedars, evergreen oak, cork oaks, and elms. The main area of flower garden lay to the east of the house: 'In ordinary cases, the front of the house is considered the right position for this, but here, with the park-like continuation of the lawn, stiff formal beds with their glare of colour would have been altogether obtrusive and out of character.'[55]

This is a sentiment with which Nesfield would almost certainly have disagreed. The beds were not intricate, 'the object being to introduce large bold areas of colour, with plenty of grass between'. Beyond the flower garden was a rosery, with beds dug 3 or 4 feet deep because of the poor, dry soil. These were 'thinly planted with standards varying in height from the centre to the sides, and the spaces between these are filled with dwarfs to cover the soil'. The rosery was enclosed by walls of different heights, against which were grown shrubs and 'specimen plants of Cupressus macrocarpa, Cedrus deodar, Arborvitae, variegated hollies' and others.

Another garden which received particular praise from the contemporary gardening press was Hardwick. Some alterations were made to the grounds in the late 1830s, but further changes were effected in the mid-1840s, and their history is recorded in a series of letters between the owner, the Rev. Sir Thomas Gery Cullum; an amateur gentleman designer called Sir John Nasmyth; and William Habersham, a nurseryman based at St Neots.[56] In 1845 Cullum approached Nasmyth for designs for new *parterres*. A plan was duly produced, and sent to Habersham to be 'worked up'. He spent a week at Hardwick and sent his proposals to Nasmyth, with the comment:

> I have made my perspective drawing of the terrace from the rosarium …
> I am delighted with your beautiful *parterre*, the chaste simplicity and bold graceful outline is striking. I am afraid I have not given the trees overshadowing the terrace rightly but the positions are correct.

Nasmyth in turn sent the completed scheme to Sir Thomas but, for reasons not entirely apparent in the sources, he refused to pay either of the two men. Nasmyth subsequently suggested that Sir Thomas's gardener should lay out the design, but not unreasonably insisted that the work already carried out must be paid for. The result of this dispute is unclear: certainly, later accounts suggest that the head gardener, D. T. Fish, laid out a formal garden here with 'very large beds cut out of turf … filled with gay plants'. By the 1880s, there was also an extensive arboretum and pinetum, an avenue of lime and sycamore, arbours of climbers, a yew walk and a dell garden or 'hardy fernery', made in an old chalk pit.[57] A visitor in 1869 noted an impressive array of glass houses, including a winter garden 40 feet long and 16 feet wide, and a conservatory/orangery, 50 feet by 17 feet 6 inches.[58] As at Shrubland Hall, the owner's wife was actively involved in the development of the gardens; an obituary, written on her death in 1875, noted that she 'was always adding to and improving the grounds'.[59]

Walks through woodland and shrubberies, often terminating at kitchen gardens, tended to become even more lengthy and elaborate in this period. At Hoxne, fragments of paths and stone steps survive amidst the outgrown shrubberies, incorporated now in a private garden; the hall itself has long since disappeared.

Parkland

Although gardens and pleasure grounds increased steadily in importance in this period, parks were not neglected. Indeed, their continued significance largely explains why the *parterres* designed by Nesfield and others were placed not in walled enclosures as the historical models on which they were based had generally been but on wide, balustrade terraces which offered unrestricted views across the wider grounds. Indeed, the layout and planting of parks continued to develop in the High Victorian period, and although fewer new parks appeared than in the previous period discussed [60] many established sites expanded in size.

At Shrubland, for example, the Middletons were not only concerned with the gardens and pleasure grounds in the vicinity of the hall; many alterations were made to the wider landscape. The most significant were in the east of the park, where the Ipswich–Coddenham road, which had previously formed its eastern boundary, was diverted by a Quarter Sessions Order of 1844. The long, straight section of the present road along Sandy Hill was thus created, and the park was expanded by a further 30 hectares. This was a slightly gratuitous extension, one might think, to what was already a vast park, but one evidently thought necessary to give the hall additional privacy. There were other alterations. Additional planting was made in various places; the prospect tower on Tower Hill was given an extra storey in 1857 (thus removing Repton's pinnacles);[61] and, most importantly of all, several new lodges were erected at the principal entrances to the park. The southern or Barham Lodge was designed by the architect Alexander Roos in a suitably Italianate style and built in 1841. The north-western lodge (Needham Lodge) was – to judge from the evidence of the Tithe Award map – already in place by 1840 but was either rebuilt or extensively altered in the 1860s, again in a suitably Italianate manner, apparently to designs by Barry.[62] The other lodges were built at various times in the middle decades of the nineteenth century. Around the same time the planting beside the principal drives was considerably embellished. By 1867, according to an article in *The Gardener's Chronicle*, the north-western drive ran

> through a most interesting wood of Oak, Beech, Spruce, Scots Firs etc. while here and there, and almost everywhere, choice specimen Conifers look out and up in cosy places. Among them were noticed Abies Morinda, Picea Pinsapo, and Pines macrocarpa. Parts of this fine wood are planted in the mixed style; other portions of it are grouped. Masses of Scots Firs, planted on bold mounds, have a fine effect.[63]

At Woolverstone, similarly, Nesfield's expansion of the pleasure grounds and gardens was accompanied by substantial changes to the park. The landscape became more wooded, lodges were erected at the main entrances, and a new more impressive drive was laid out, providing a suitably imposing approach from the direction of Ipswich. It led through a great western extension to the park laid out in the 1850s. This is now a distinct property called Freston Park, physically separate from Woolverstone and under different ownership. Its history is complex and of some interest. The manor of Freston passed to the Latimer or Latymer family in the early sixteenth century, and it was they who, some time around 1560, built the romantic prospect tower which survives beside the site of their manor house here. It is an imposing brick structure which testifies to the fact that even in the sixteenth century the view across the Orwell estuary was considered an appealing one. By the early eighteenth century the site had declined in status and was known as Hall Farm or Tower Farm. Some time before 1795 the estate was acquired by Charles Berners of Woolverstone, and a map of that date shows the hall and tower in the centre of an agricultural landscape of enclosed fields, with areas of alder carr and marsh along the banks of the River Orwell.[64] Soon after 1850 a substantial part of the farm was laid to grass, the hedges all removed and many new trees planted. The new western drive, a mile and a half long, ran through it from end to end. It led to a new lodge, the Monkey Lodge, on the Ipswich Road. The drive was lined with copper beech trees, many of which still survive, and offered striking views across the picturesque Orwell estuary, as well as fine prospects towards the Freston Tower.

At Somerleyton, likewise, the elaboration of gardens by Nesfield was accompanied by an expansion of the park. When the estate was acquired by Morton Peto in 1846 the park was not large. The sales particulars suggest that it covered 120 acres (48 hectares) but the attached map suggests that only around half of this formed a designed core, with the rest made up simply of adjacent pasture fields.[65] By 1861, however, when the estate was once again placed on the market, the park had more than doubled in size and was an altogether more impressive affair:

> Lying in the most compact form, opening conveniently to the High Roads and intersected by a fine carriage drive in excellent order [it] contains altogether 260 acres 1 rood 19 perches of which 60 acres 3r 24p are ornamental plantations and fine pheasant preserves, 197a or 21p fine pasture on a deep loam and well timbered around and in sight of the mansion grounds with two noble double avenues of Limes and Elms.[66]

Peto seems to have lost little time in preparing for this expansion. In 1848 a Quarter Sessions road closure order terminated and diverted a number of public rights of way in the area to the south and west of the park.[67] Not only was the park immediately expanded in this direction but in addition a diminutive park which had long existed around the neighbouring Somerleyton Rectory was now united with it to create what was in effect a single designed landscape. New carriage drives were laid out. The main one led northwards from the new public road along the southern boundary of the park. A lodge was erected here, 'brick with Caen stone dressings, with Ornamental Tiles, Brick and Stone Panel-wings to the Iron gates'. Another led in from the west, and this entrance too was supplied with a lodge, which in 1861 served as home for the game keeper: 'rustic, brick, rough cast with reeded roof ... very ornamental'.[68] The offices included 'a detached rustic slaughter house'. As at other great houses, these lodges both provided a measure of security and privacy, and also advertised the presence of a residence of importance; the long wall of brick, which was now built almost all the way around the perimeter of the park, did likewise.

Most of the trees growing today within the area of the expanded park are oaks; many of these were, needless to say, former hedgerow trees. But a significant number of beeches were also planted, some of which survive, and it is clear from map evidence that conifers, probably pines, were also widely established in the park at this time. Only a few remain, but I have noted before how the existing parkland timber often gives a poor guide to the character of the original planting. Peto's planting also featured a number of ornamental clumps, largely dominated by oak although some contained horse chestnut, beech, copper beech and lime in various combinations, as well as conifers. The planting was carefully considered: the most complex and ornamental clumps were located close to the hall or beside the principal entrance drives – like the fine example, comprising six limes and one purple beech which stands to the south-east of the pleasure grounds.

New belts and plantations were established along the western, north-western, and south-eastern boundaries to the expanded park. These, too, are of some interest. The southern belt was, to judge from the surviving trees, largely dominated by oak, although examples of *Pinus nigra* and *Cedrus libani*, together with an understorey of yew and holly, suggest a desire to provide year-long colour – the woodland formed a backdrop to the views within the new, southern extension to the park. The north-west belt was a more complex affair. The 1861 sales particulars describe it as 'West Park Shrubbery', and it evidently formed an extension of the pleasure grounds around the hall. The growth pattern of the older trees found in this area today – with many low branches – indicates that they originally grew in a fairly open environment: the species present include beech, copper beech, oak, lime, plane, cedar, pine, yew, Irish yew, holly, *Thuja picata* and Portugal laurel. The western belt contains a similar range, although the more ornamental species only survive well towards its southern end. While oak, sweet chestnut and beech form the bulk of the planting, cork oak, yew, Irish yew, various forms of holly, rhododendron, privet, box and Portugal laurel are also present. The growth pattern of these trees suggests that this area was originally more densely planted than the north-western belt. The map which accompanies the sale particulars shows, once again, that the area was threaded with paths; the box which grows here in some abundance may once have lined these.

Victorian gardens

The paths led to the village of Somerleyton, largely rebuilt as a model settlement by Peto to designs by John Thomas (Figure 47). This is unquestionably one of the finest examples of a picturesque estate village in the country, comparable to Blaise Hamlet near Bristol. The ornate cottages with twisted 'Tudor' chimneys, decorative barge boards and some mock timbering cluster around a village green. Ironically, the settlement was erected on what had once been Somerleyton common, enclosed in 1803 – the older properties in the village survive, lying some way back from the new Peto additions, along the line of the old common edge. An article in the *Florist* in 1857 was positively gushing about the village, emphasizing

> How considerably the comforts and conveniences of the inmates had been provided for by the in-door arrangements of the various styles of cottages, as in their well-kept gardens where the *useful* was not altogether sacrificed to the *ornamental*, for we noticed an ample supply of vegetables in each, besides a gay assortment of showy flowers, and the walls were crowded with masses of gay Roses, Honeysuckles, and Jasmines.[69]

FIGURE 47. The picturesque estate village at Somerleyton formed the termination of walks leading out from the hall.

The village was indeed a very public advertisement of Peto's paternalism. It was provided with a shop and even a police house, but not a pub; Peto was

a staunch Baptist, although he did rebuild the parish church in 1854, which now stood isolated within the park as a result of the recent expansion.

Many of the developments at Shrubland and Somerleyton were mirrored (albeit generally on a smaller scale) at other places. Expansion was common: the park at Benhall for example grew from 49 to 70 hectares between 1836 and 1847,[70] while that around Hardwick House, which covered *c.* 30 hectares in 1837, had by 1892 expanded to both east and west, and covered an area of around 47 hectares. Branches, Dalham, Thornham and many other parks experienced significant expansion. Ornamental buildings continued to appear, like the diminutive gothic folly at Thornham, probably erected in the 1860s, or the gloomy mausoleum erected by the Barretts in Brandon Park in the late 1860s. As at Somerleton, walls were sporadically built around part, or occasionally all, of the perimeter, as at Great Saxham. The most astonishing example must be that at Easton, where the wall which runs all around the park (a distance of 3 kilometres) is for most of its length of 'crinkle-crankle' or serpentine form; it is the longest such wall in the county and one of the longest in England. It is, unfortunately, in poor condition: being only one brick in width, it has now collapsed along much of its length, although some well-preserved stretches survive along the western boundary of the park.

Lodges increased steadily in number, and entrance drives were generally more carefully and elaborately planted than in the landscape parks of the eighteenth century. That at Orwell Park, for example, leading off from the Ipswich Road, 'Winds through ornamental grounds planted with Conifers and groups of tall Limes and Elms, until it reaches an avenue of Araucarias and *Pinus insignis*, planted alternately.' No trace of this avenue now survives. It may not have lasted long after this description was made, for *The Gardener's Chronicle* went on to report that 'owing to the light nature of the soil and the extreme dryness of the climate in this locality, neither the Araucaria nor the Pinus succeed satisfactorily.'[71] Better preserved is the rather classy treatment of the northern and south-western entrance drives at Sotterley: the former takes the form of a Douglas fir avenue with flanking clumps of Wellingtonias, all probably planted in the 1860s; the latter is a curving, and irregularly planted avenue of horse chestnuts interspersed with lime, perhaps a little earlier in date.

Almost all Suffolk parks, even the oldest, are dominated by nineteenth-century planting. Oak continued to be the favoured tree – or at least, that which has survived the rigours of the local environment with the greatest success – but beech was widely planted on the lighter soils (especially in clumps and belts, as at Elveden), and horse chestnut, lime, and sweet chestnut are well represented (as for example at Benhall). Conifers were now more popular than ever, both in specialised arboreta and within the wider parkland, as single trees or in clumps. When Nowton Court was rebuilt by J. H. P. Oakes in 1875 and the diminutive landscape greatly expanded, a particularly varied collection of trees was established. Mature trees were moved and planted, 'upwards of 40′ high', using one of William Barron's machines. *The Gardener's Chronicle* for 1883 refers to:

Picea Morinda, Pinus austriaca, Abies Douglasii, scarlet Chestnuts, ever-gren oaks, Yews, Thuja gigantea, Libcedrus decurrens … Cedrus libani and atlantica, Wellingtonias, Txodium sempervirens, Cryptomeria japonica and elgans, Thulopsis borealis, and variegated Retinosporas.

The free-standing conifers were pruned so that the lower bole was bare, giving the effect of the 'grazing line' apparent on deciduous parkland trees.[72] The new park laid out at Bradfield Combust in the late 1850s was likewise planted by its owner, A. J. Young, with a remarkable collection of conifers, some of which still survive, including Cedar of Lebanon, *Pinus nigra*, Atlas cedar and Wellingtonia.

Although, as we have seen, exotic conifers are often less hardy and shorter-lived than native trees, many Suffolk parks still contain examples of mid-late nineteenth-century date: fine Wellingtonia can be seen at Campsea Ash, Hengrave, and Benhall; Cedar of Lebanon (now widely planted in the open parkland) at Campsea Ash and Hengrave. Particularly striking examples of nineteenth-century coniferous parkland planting exist at Or-well Park (mainly *Pinus Wallichiana*, *Pinus nigra*, *Pseudotsuga Menziesii*, and *Pinus radiata*) and at Elveden (Scots pine, Cedar of Lebanon and Wellingtonia).

The paths at Somerleyton, which led through woods and shrubberies to the neighbouring village, were part of a wider pattern. Following the lead set by Repton and others at the end of the previous century, ornamented walks were increasingly established, leading out from the formal garden areas into the parks and plantations. At Orwell Park an extended walk began in the shrubbery, bounded by:

> Banks of common laurel, which are kept neatly cut, and, to relieve the monotony the trimmed surfaces would otherwise present, specimen plants of Portugal Laurel, Yew and Holly are allowed to grow out. Standard scarlet Thorns and Laburnums are also planted at a suitable distance.[73]

The path gradually changed 'more into the wild character' and then entered the park, leading ultimately to a boathouse by the Orwell estuary, 'thus forming nearly a mile of delightfully shady walk or drive, bordered on each side by flowering shrubs or wild flowers'.[74]

The return of the garden to a place of prominence in the country house landscape, and the revival of archaic forms of *parterre* by Nesfield and others, was part of a general upsurge of interest in the distant past which was also manifested in the adoption of gothic or Elizabethan styles for some new houses in Suffolk at this time, such as Flixton Hall, systematically rebuilt by Sir Robert Shafto Adair in the 1840s (and given, as we have seen, fashionable gardens by Nesfield). Antiquarian revival also influenced the ways in which surviving relics of old, geometric landscaping were regarded. In particular, where avenues still existed these were now more likely to be retained and replanted than removed. Thus the ancient lime avenues at Campsea Ash appear

to have been comprehensively replanted in the mid-nineteenth century, as were those at Kentwell, Rougham, Broke Hall at Nacton, and Hengrave. In the latter case, while lime was used for the main body of the avenue, the southern end was planted with Cedar of Lebanon, a curious but visually striking combination (Figure 48). Deer continued to be maintained in many parks, although now almost entirely for the antiquarian kudos they provided. In 1848, when the estate at Orwell was sold to George Tomline, a separate sum of £3,000 was paid, 'being the price of the deer, Garden Implements, Appendages, Greenhouse Plants, and other articles and things belonging to Sir Robert Hartland, Baronet'.[75]

This book has largely been concerned with the great designed landscapes laid out around the homes of large landowners, rather than with the more modest grounds created by those further down the social scale. In part this is because, in the period before the nineteenth century, we have comparatively little information about the character of smaller gardens; in part it is because they have left few tangible traces in the modern landscape. This begins to change in the middle decades of the nineteenth century, however. Many middle-class residences in both town and country contain some planting from this period, especially of specimen trees – copper beech, Wellingtonia and Cedar of Lebanon appear to have been especially popular.

Rectors and vicars were particularly enthusiastic gardeners, and the gardens of many rectories and vicarages still boast fine collections of trees. One of the

FIGURE 48. The avenue at Hengrave Hall was systematically replanted in the nineteenth century, partly with lime, partly with Cedar of Lebanon.

most astonishing examples was laid out by George Drury, Rector of Claydon, in the 1850s and 1860s. Local tradition holds that it is an allegorical, biblical landscape. In the area immediately to the south of the churchyard – which abuts on the rectory garden – is a substantial walled garden, with decorative gateways and two three-storey flint towers incorporating medieval masonry taken from the chancel of the adjacent church, which was demolished and rebuilt around 1860 by Drury (Figure 49). Nearby there was, until recently, a shell grotto, and further to the south a small pond and 'river' (now dry), crossed by a bridge. The walled garden is said to represent Jerusalem, and one of the towers the house of the last supper; the grotto is perhaps the sepulchre of Christ; the 'river' the Jordan. Some credence is given to this interpretation by the character of the Rev. Drury: he was a Tractarian, and a leading figure in the Oxford Movement, who founded a convent in the village. The nature of the planting, which survives in neglected and degraded form, choked with invasive sycamore, is also suggestive; it includes, or included, examples of Tree of Heaven (*Ailanthus altimissa*), Judas Tree (*Cercis siliquastrum*), and Christ's Thorn (*Paliurus spina-Christi*), the latter a particularly rare tree.

FIGURE 49. Claydon Rectory: one of the folly towers built by the Rev. George Drury.

143

CHAPTER 7

The business of gardening

Gardeners and nurserymen

In all periods, as we have seen, landowners and their families often took a keen interest in the practical management and organisation of their gardens, delighting in the production of flowers, vegetables and fruit. A particular enthusiasm for gardening seems to have run in certain families, like the Hanmers of Mildenhall or the Cullums of Hardwick and Hawstead, as Pat Murrell's extensive research has demonstrated. We have seen Sir Dudley Cullum in the 1680s busy with the collection and nurturing of plants, and the Rev. John Cullum in the 1780s carefully noting the growth of his trees and shrubs in the gardens at Hardwick. For the years around 1800, Sir Thomas Gery Cullum's Commonplace Book contains many notes and musings on matters horticultural, including ideas about how to preserve apples over the winter, and how to make concoctions for curing 'diseases in trees' and 'to stop the bleeding of Vines and other Trees after pruning' (a mixture of cheese and oyster shells!).[1] But the interests and enthusiams of such people could not have been sustained without the knowledge and experience of professional gardeners and the hard work of undergardeners and labourers, or in the absence of a commercial infrastructure of nurseries, seedsmen and suppliers of gardening equipment. This essential business of gardening became more complex and sophisticated in the course of the eighteenth and nineteenth centuries.

Problems and ambiguities of nomenclature make it hard to chart the development of commercial nurseries and seedsmen in Suffolk in the seventeenth and eighteenth centuries. The term 'gardener' in wills, leases or trade directories could have a variety of meanings, ranging from market gardener, through seedsman and nurseryman to jobbing labourer. Thus the precise status of men like John Norris, 'gardener', who was party to deeds in Bramford in 1663, or William Chapman, 'gardener' of Ipswich, who made his will in 1738, remains obscure.[2] Nevertheless, it is evident that commercial suppliers of seeds and plants already existed in Suffolk by the early years of the eighteenth century. In 1734, for example, Christopher Winterflood was advertising a variety of trees, including walnuts, in the *Suffolk Mercury or Bury Post*.[3] As early as 1709 an inventory from Little Thurlow Hall described various plants and vegetables obtained 'from the seed shop'.[4] One particularly long-lived nursery was that set up by William Woods at Woodbridge in 1749. It remained in the family throughout the eighteenth century (John and William Woods were county freeholders there in 1798)[5] and into the nineteenth – a series of 'John Woods', variously styled gardener, seedsman and nurseryman, of Cumberland Street, appear in White's Directory from 1844 until the 1890s. By 1900, however, the premises had been taken over by a new proprietor, Roger C. Notcutt. His firm became, of course, the most successful nursery and gardening business in the county.

Equally long-lived was the nursery at Long Melford established some time before 1750 by Timothy Constable. On his death in 1751 the business was taken over by his widow Elizabeth and brother Thomas.[6] The company was still operating in 1798, now under the charge of another Timothy Constable. The other main firm operating in the county by this time was Coe's in Bury St Edmunds, which placed frequent advertisements in the local papers.[7] In the *Bury and Norwich Post* for 13 January 1802, for example, Jarvis Coe advertised that he had

> Left the Swan Inn, Northgate-Street and is removed to the next House, where he hopes to have a continuance of that patronage which has ever given him the most heartfelt satisfaction. Having now no other employment, he resolves that his chief study and delight shall be to bring every Plant and Shrub in his Garden to the highest perfection, and to shew what Nature can produce when assisted by art and experience.[8]

He boasted of his 'choice collection of Hot-house and Green-house Plants, seeds of all kinds, Flowers, Fruit Trees, Shrubs, and Herbaceous Plants, with every other article in the gardening line' (an advertisement in June of the previous year gave his reasons for leaving the Swan as a 'consequence of not agreeing to sell Common Brewers Beer').[9] The business was a well-established one, for in 1802 T. Cullum described in his Commonplace Book the great vine 'of the black Muscadine kind' planted in the garden here ('opposite the Grammar School') about fifty years earlier by James Coe, Jarvis's father. It extended over 150 foot of wall – one branch alone was 79 feet long.[10] It was still extant in 1810, when William Hervey reported that 'its branches extend at present 100 feet; he [Coe] lately cut off 40 feet from the extent …'[11]

The nineteenth century saw an increase in the number of suppliers, although again ambiguity of terms makes it hard to chart this with precision. White's Directory for 1844 for example lists thirty-six individuals in Ipswich as 'Gardeners, Seedsmen etc.', but some of the entries are qualified with 'florist' and only one with the word 'nurseryman'. This was the firm run by Robert and James Jefferies of Corn Hill and London Road.[12] The two brothers later took a lease on the arboretum ground on Henley Road in St Margaret's from the Mayor and Aldermen of the town.[13] In 1864 White's Directory employed a different terminology, describing six individuals in the town as 'Nurseryman, Seedsman and Florist', but there was still only one – Jefferies – clearly described as one of the former.[14] Ten years later, however, no less than six individuals were specifically described by the Directory as 'nurseryman' in the town, compared with two in Bury St Edmunds.[15]

Even with this increase, however, it is apparent that in the eighteenth and nineteenth centuries Suffolk rather lagged behind neighbouring counties in the provision of nurseries and seedsmen. Suffolk has nothing to compare with the great nursery established by John Aram, on the outskirts of Norwich in the 1750s – later known as Aram and Mackie's – which supplied most of the major gardens in Norfolk, and several in Suffolk (such as Benhall, in the 1820s).[16] In part, this may have been a consequence of the county's relative proximity to London; Norwich, where Aram and Mackie's flourished so successfully, was just that bit further away. Certainly, Suffolk estates often purchased plants directly from the capital. In the late seventeenth century, when Sir Dudley Cullum was sent plants from George Ricketts of Hoxton, near London, this was because local suppliers simply did not exist.[17] But even at the end of the eighteenth century London firms were prominent in the local market place. In 1797, for example, the *Bury and Norwich Post* carried a lengthy advertisement from Davies' of Berkeley Square in London, for 'a new sort of celery' from the island of Samos, 'Lettuce-Leaved African Endive', and 'Tuscan Branching Purple Brocoli ('it has the richest flavour possible, and its beauty when on the table must be seen to be believed').[18] In 1847 Lord Hervey was purchasing plants for Ickworth direct from the nursery of John and Charles Lee in Hammersmith,[19] while in the 1860s the Gages at Hengrave were purchasing plants from the London nurseries of Lee, Veitch and Waterer,[20] a firms which also supplied most of the plants when the gardens at Boulge Hall were remodelled in the 1890s.[21] The fact that provincial nurseries outside Suffolk often advertised in the local press implies that local suppliers were poorly developed. In December 1801 S. Ellington of Brandon advertised 100 walnut trees 'of that celebrated sort, called Hall's High-Flyers';[22] in March of the following year seeds of the Brompton Stock were advertised for sale by 'Mr White' of Wisbech and 'Mr Newcombe' of Stamford.[23] It is noteworthy that when the *parterres* at Hardwick were being laid out in 1845, Rev. Sir Thomas Gery Cullum called in not some local firm but William Habersham, a nurseryman based at St Neots.[24]

Nevertheless, if local nurseries were perhaps less developed than in some adjacent counties, advertisements in the newspapers in the eighteenth and nineteenth centuries attest the commercial importance of gardening: all manner of things were advertised, bought and sold. There was even an active trade in second-hand plants: in April 1802 'A peculiarly well-chosen COLLECTION (comprising about 300 Pots) of AURICULAS, unrivalled for excellence of sorts, having been selected from the first catalogues in the kingdom' were offered for sale by auction at Hasketon.[25] Moreover, Suffolk played a key role in the development of garden technology. Ransomes, the Ipswich company of agricultural engineers founded by Robert Ransome in *c.* 1780, began to produce lawnmowers as early as the 1830s, having purchased the right to manufacture Edwin Budding's patent design in 1832. By 1852 more than 1,500 machines had been produced, and in 1872 the company built a separate works primarily devoted to lawnmower manufacture. With the onset of the agricultural depression in the 1880s, this facet of Ransomes' business became particularly important.[26]

Once trees and plants had been purchased, they required the care of trained gardeners. In the early eighteenth century garden staff appear to have been limited in number and even large estates often employed only one principal gardener, assisted by labourers who were employed for much of the time on other tasks. Such gardeners seem to have enjoyed some prestige, and they often moved from estate to estate: men like James Taite, who was principal gardener on the Marquis Cornwallis's estate at Culford in the 1790s, but who had previously worked for Sir Charles Kent at nearby Fornham St Genevieve, and before that had been employed at Livermere by Nathaniel Lee Acton.[27] The pay of such men compared favourably with that of other senior servants: even on a small estate like Wortham the gardener received £16 per annum in 1775, a sum which had risen to £25 by 1811.[28] As gardens grew more extensive and complex in the course of the nineteenth century, however, and garden staff grew larger and more specialised, the status and pay of head gardeners grew accordingly, especially on the larger estates. At Benhall, estate accounts from the 1820s and 1830s describe the work of three or four men regularly employed in the garden: 'trimming trees', 'painting lights', 'hoeing round gardens' and 'weeding'.[29] But at places like Shrubland, where gardens of particular magnificence were created, many more men were needed. By 1865 a head gardener, three under gardeners, twenty-five labourers and seven boys, as well as a plumber, a painter, a carpenter and a carpenter's boy were all paid on the garden account.[30] Even when agricultural depression set in towards the end of the century, garden staff remained numerous on some estates. At Elveden in the 1890s there were twenty flower and twenty vegetable gardeners;[31] while an account of the gardens at Livermere drawn up in 1899 describes:

> Area of Kitchen Gardens 14 acres (including 1½ acres of water). Area of Pleasure Grounds, including Hall, 11 acres. Number of gardeners employed: 1 Head Gardener, 5 able-bodied men, 1 old man, and 2 boys.[32]

Head gardeners on the largest estates not only commanded large salaries – £100 per annum at Shrubland Hall by 1865 – but they also often enjoyed celebrity status. Men like Donald Beaton or D. T. Fish at Hardwick went on to become regular writers in the burgeoning gardening press. Conditions and accommodation improved for such men in the course of the nineteenth century. *The Gardener's Chronicle* for 1876, describing a visit to Orwell park, noted that:

> The gardener's house is pleasantly situated a short distance from the gardens, and it is gratifying to find that a class of men who minister so much to the enjoyments and pleasures of the wealthy are of late being more studied in respect to their dwellings than formerly, and are altogether attaining a better status, more in accordance with their merits.[33]

But the trainees and undergardeners were offered less elaborate accommodation, sometimes in small bothies built against the north wall of the kitchen garden. Increasing numbers of staff, and the rising status of (and competition for) accomplished gardeners, ensured that wage bills escalated through the middle years of the century. At Shrubland in the first six months of 1865 £559 9s. 9d. out of a total maintenance bill of £913 19s. went on labour costs, although few other estates would have come close to this.[34]

Kitchen gardens – design and location

We have been largely concerned in the previous chapters with ornamental, aesthetic gardening – with parks, *parterres* and pleasure grounds. But throughout the eighteenth and nineteenth centuries all Suffolk landed estates also maintained kitchen grounds and orchards which supplied the household with fresh food. In the seventeenth century the kitchen garden had simply been one among the collection of enclosures clustering around the mansion, along with the stable yard, entrance court, nut grounds, ornamental gardens and the rest. Where, as occasionally happened, owners resisted the tide of fashion and maintained at least some of the walled enclosures around their homes, the kitchen garden stayed where it was. Indeed, even where some of the domestic complex was comprehensively rebuilt at a later stage, the location of the kitchen grounds often displayed remarkable stability. At Dalham, for example, the kitchen garden occupies the same area today as it did when a survey of the hall and its surroundings was made in 1728.[35] More usually, as we have seen, the walled enclosures were entirely removed from the vicinity of the mansion in the course of the eighteenth century, as the fashion grew for setting the mansion 'free of walls' in open parkland. When this occurred, the kitchen garden was likewise moved, but there is some misunderstanding about its destination, and about what this change signified.

This misunderstanding has a long history. William Andrews Nesfield, writing about Henham in the 1850s and looking back to the previous century, described how 'Kitchen gardens were deemed nuisances & therefore were placed in most instances at distances varying from ¼ to 1 mile.'[36] But this is, to put it mildly, an exaggeration. There are some Suffolk examples of kitchen grounds relocated in this period to some distant recess of the park, but they are rare. Even a cursory examination of contemporary plans shows that in most cases the kitchen garden, while hidden from view from the park, remained fairly close to the house. In some instances, it was tucked away behind the stables – a location which allowed manure to be transported to it, without giving offence to the family and their visitors. But in many cases it was also positioned close to the pleasure grounds, and could often be accessed directly from them. Indeed, Capability Brown's own designs for Heveningham and Branches show such an arrangement, the house, stables, pleasure ground and walled garden in both cases occupying an ovoid area divided from the park by a ha ha.[37] Contemporary maps frequently show that paths winding through the pleasure ground led to the walled garden: these areas were supposed to be visited by members of the family and their friends. Some, at least, seem to have contained flowers and, on a cold spring day, made a pleasant place to take the air. Repton wrote in praise of the recreational potential of the kitchen garden – perhaps with memories of his native Suffolk in mind: 'There are many days in winter when a warm, dry but secluded walk, under the shelter of a north or east wall, would be preferred to the most beautiful but exposed landscape; and in the early spring … some early flowers and vegetables may cheer the sight.'[38] Yet it remained imperative that the garden should be invisible from the park: nothing should detract from the impression that the mansion stood alone, free from all necessary and useful facilities. To achieve this, however, the removal of the kitchen garden to some distant place was not generally necessary. As one early nineteenth-century writer put it, 'it can seldom happen that the garden walls may not be effectually concealed by means of shrubs or low-growing trees so as not to be seen.'[39] To judge from the surviving remains of such screens, yew and box were particularly favoured for this purpose, providing as they did good year-round cover. Today, many kitchen gardens *are* visible from the parkland, but this is only because of the neglect and gradual disappearance of these kinds of shrubbery screen.

There are, of course, some examples in Suffolk of kitchen gardens located, in the way Nesfield described, at some distance. But in several instances this was because the mansion had been rebuilt on a new site, leaving the old gardens where they were. At Shrubland Hall, for example, the kitchen grounds laid out around the Old Hall remained in use long after the construction of Paine's new hall, some 500 metres to the south, in the 1770s – a single apple, tree still grows in the parkland marking their site. At Ickworth the kitchen garden lies some 400 metres to the south of the house. Here the garden originally stood beside Ickworth Manor, which was demolished in 1710, the Earl of Bristol intending to build a new house on the same site. The walled garden, like the canal constructed beside it, the summer house and the orchard, were originally planned to go with this new house.[40] The Hervey family moved to Ickworth Lodge on the north side of the park but the new house was not built until much later – in the 1790s – and then on a different site again. The walled garden remained where it was, but was not neglected – with its fine canal, made more serpentine in the later eighteenth century, and elegant summer house, it formed the termination for a well-trodden path from the mansion house, past the parish church (itself suitably 'landscaped'). Similar again is the case of Stowlangtoft, where the kitchen garden today lies in the park some 400 metres to the south of the hall, surrounded by plantations. Until *c.* 1860, however, kitchen garden and hall lay close together: once again, isolation resulted from rebuilding on a new site.[41] Indeed, only a handful of country houses in Suffolk seem to have had kitchen gardens which were, by design and from the start, located at any distance from the house – as at Livermere, where they lay nearly a kilometre to the east of the mansion.

It is noteworthy that even here, to judge from the Tithe Award map of 1847, the garden was reached by a path from the hall, and had an area of detached pleasure garden associated with it.[42] This was probably created by

Repton. In the Red Book for Livermere he complimented the 'beautiful scenery surrounding the Kitchen Garden', and proposed a new path leading to it, embellished with ornamental planting and a 'rude bench'. Nevertheless, the walled enclosures needed to be kept hidden from the surrounding park. The fruit garden here was to be 'entirely walled up, and shut out from the surrounding scenery; I must recommend colouring the red wall with a dark wash, where there is not sufficient depth to plant it out entirely'.[43] Repton suggested adding or improving walks to the isolated gardens at Shrubland, as we have seen, and at Henham.[44] Similar paths can be detected at most places where a fair degree of separation between house and kitchen garden was achieved, as at Hoxne, where the garden lay over 200 metres south-east of the house. The intervening area was occupied, at least by the early nineteenth century, by areas of shrubbery, pleasure ground and garden.[45]

In the middle and later decades of the nineteenth century the kitchen garden was integrated even more closely with the pleasure grounds and ornamental gardens. The arrangement already described at Somerleyton (above, pp. 130) – where paths led directly from the house, through the middle of the pleasure grounds on the north lawn, to an ornamental entrance to the walled garden – could be paralleled at many other sites. Even where maps appear to show the garden shunned and removed to some distant location, contemporary descriptions reveal a more complicated picture. At Orwell Park, for example, the garden actually lay outside the park altogether, but an article in *The Gardener's Chronicle* for 1876 described how:

A broad gravel walk winds along one side of the rosery, and leads from the lawn to the kitchen garden, which is reached by crossing the public road. The sides of this walk are planted at equal distances from the edges, with tall pillar roses and upright Irish Yew alternately.[46]

Perhaps not surprisingly, an examination of the surviving structure of eighteenth- and early nineteenth-century kitchen gardens usually reveals that considerable care and attention was paid to their form and layout. As Marilyn Williams has shown in her important study of Suffolk kitchen gardens, various attempts were made to maximise the length of warm, south-facing wall (on which peaches and nectarines were normally grown).[47] At Branches, Brettenham, Broke Hall, Coldham, Orwell Park and Rougham, the main enclosure was kite-shaped (with the angles directed towards the compass-points), a shape recommended by Loudon in his *Encyclopaedia of Gardening*: 'a square set out in such a manner that each wall may have as much benefit of the sun as possible'.[48] At Great Glemham (Plates 18, 19) the north wall is polygonal, almost curving, while at Boulge, Culford, Ickworth and Melford the garden enclosures were trapezoid – almost like triangles, but with the northern corner cut off. Perhaps the most striking plan was that of the new gardens erected in the park at Shrubland Hall in the 1840s. An area of 1.6 hectares was contained within an eight-sided enclosure, essentially an octagon but with longer sides running east–west. Even some of the more irregular shapes occasionally encountered were usually the result of careful consideration rather than lack of interest or concern. The strange shape of the enclosure at Hoxne, for example, is partly dictated by topography (the ground falls away steeply to the south) and partly by the unusual layout of the south wall, which bows inwards in an elegant curve – once again, to maximise the length of south-facing surface.

Interesting though such examples are, they were always in a minority. The vast majority of kitchen gardens built in Suffolk in the eighteenth and nineteenth centuries – more than three-quarters, according to Marilyn Williams – were of square or rectangular shape.[49] Most contemporary writers advocated such a plan as the most useful and practical. Many however were subdivided, often some time after they were first erected, again in an attempt to increase the length of south-facing walls. At Heveningham, unusually, an east–west crinkle-crankle wall was added to the existing enclosure (probably of *c.* 1720) in 1796, and still survives in good condition. At Parham a crinkle-crankle wall of similar age surrounds the whole of the garden, covering about one-third of an acre (other examples survive at Coney Weston Hall and Melford Hall). The wavy shape was supposed to provide extra stability to the wall, and give shelter to trees growing in the alcoves.

Writers on gardens and gardening in the eighteenth and early nineteenth centuries, such as Abercrombie or Loudon, recommended that – ideally – the longest sides of the garden enclosure should be ranged east-west, and that the garden itself should occupy a gentle south-facing slope. Loudon, for example, stated that there was a 'general agreement to prefer a gentle declivity towards the south, inclining a little towards the east to benefit from the morning sun'.[50] The builders of Suffolk's kitchen gardens often followed the first piece of advice, but their adherence to the second was more patchy, determined by the character of the local topography. Even the first was sometimes ignored, often because they were dealing with sites which were partly or entirely moated: examples include Livermere, Otley, Crows Hall, Gedding and Kentwell. In most of these cases there is little doubt that the kitchen garden occupies an area already used as a garden in the fifteenth or sixteenth centuries; most of these gardens (Livermere is an obvious exception) are associated with imposing houses of fifteenth- or sixteenth-century date, which were not demolished and replaced in the eighteenth century by something more fashionable. As we have seen, many gardens of this period were probably moated, partly for aesthetic and partly for practical reasons: moats helped to drain the site (very necessary on heavy clay soils) and also provided a ready source of water. Indeed, many kitchen gardens which (as far as we know) were established *de novo* in the eighteenth century had substantial canals close at hand, as at Thornham. Moats supplied other benefits. As one writer recalled in the 1930s, looking back to childhood visits to a house by the Waveney, 'The old moat was always kept fairly clean, for each winter it was dredged for the valuable manure it yielded, much of which was devoted to the kitchen garden.'[51]

The material culture of gardening

Not surprisingly, the available evidence suggests a steady proliferation in the tools and equipment used by Suffolk gardeners in the course of the eighteenth and nineteenth centuries. Early descriptions and inventories suggest fairly modest provision. Thus at Little Thurlow in 1709 the garden equipment included a simple collection of tools – spades, rakes, 'Large shears' and a 'grass edgin ioin'[52] More varied were the contents of Sir Dudley Cullum's garden at Hawstead in 1720, which included numerous 'glasses', presumably used as cloches to raise seedlings. There were also 'Ten stock of Bees', while the store room next to the kitchen had '14 mushroom Glases with mushrooms'.[53] A century later even quite modest estates could boast a far greater range of equipment. An inventory for the garden at Langham Hall in 1832 listed:

> Iron roll – One cucumber double frame – single ditto and light – sundry sea cale pots etc. – sundry tools. 2 wheel barrows – 3 ladders 2 watering cans, shelves in chambers. Sundry old light frames – long ladder – 4 stoves in Gardener's house. Truck cart – flower stages – Flower barrow – water tub, 12 linen posts – sun dial on stone pillar.[54]

Perhaps the most important change in the course of the nineteenth century was in the provision of glasshouses. Greenhouses in the modern sense were unknown in eighteenth-century gardens. Orangeries and greenhouses – used to house half-hardy exotic evergreens – were masonry buildings with sash windows, often heated by external stoves. Examples survive at Heveningham (built to designs by Samuel Wyatt in 1778) and, until relatively recently, survived at Redgrave. Into the early nineteenth century versions of such structures were still being made: drawings for two, proposed for the gardens at Martlesham Hall in the 1820s, survive in the Ipswich Record Office.[55] By this time, however, glasshouses of modern type were appearing. Before the repeal of the glass tax in the 1840s, small panes were cheaper than large and these were generally used, slotted into closely-spaced glazing bars (such small panes were, moreover, less likely to be uneven and pitted, and therefore less prone to frost-damage, than large ones).[56]

We should not underestimate the sophistication of gardens in the early years of the nineteenth century: even fairly modest estates often had several glass houses and sometimes heated walls by this time. At Langham Hall in 1820, for example, there were 4 acres of kitchen gardens behind the stables, 'Walled in with lofty brick walls in 4 divisions, covered with choice fruit trees in full bearing, and a neat Gardener's Cottage, Conservatory, Hot-houses, Grapery, and Hot Walls for the forcing of fruit.'[57] Nevertheless, it was in the middle and later decades of the century that kitchen gardens reached their most elaborate and developed form. From the 1840s glasshouses began to take on their modern appearance, grew more sophisticated and proliferated steadily in numbers and type. By 1883 the kitchen garden at Campsea Ash, although not a large one by the standards of the time, nevertheless had an impressive range of glass.

The walls are Well clothed and the Ground stocked with Standard and other Fruit Trees. The Glass Houses comprise Green House and Stove Fernery heated by Hot water, a Conservatory, a range comprising Two Vineries and a Peach House heated by Flues. The Forcing Grounds comprise Two Houses for Gardinias and Roses, heated by Hot Water Pipes, a Twelve-light Brick Range of Forcing Pits, Cucumber Pit, Carnation House, Ranges of Marrow Pits, Two Potting Sheds, Rose Nursery, Orchard and vegetable Garden.[58]

The most important houses had even more impressive ranges, especially where – as at Somerleyton – the kitchen gardens were located beside, and formed a direct extension of, the pleasure grounds. By 1853 Crowe's *Handbook* could describe this garden in glowing terms, making it clear that it was an area of aesthetic display, not just one for vegetable production:

The vineries are very extensive, and in the centre is a light and elegant rosary. An archway at the back leads to more vineries, forcing pits &c., behind which rises an ornamental chimney, that is seen at an immense distance, being an object of equal utility and adornment, for it takes up all the smoke produced by the various fires in the different hothouses.[59]

The sales particulars of 1861 provide more detailed information.[60] The melon and forcing grounds contained two lean-to 'ridge and furrow' forcing houses and six ranges of forcing pits. The plant ground contained a range of four vineries, and a second range with a further vinery and a peach house, again with 'ridge and furrow' roofs. There were apricot and fig houses, and on the south side of the south wall (in other words, within the pleasure grounds) there were 'Glass wall houses or promenade houses, both heated'. These were a particularly elaborate version of the kind of 'hot wall' we have already encountered beside the Fountain Garden at Shrubland (above, pp. 123). They comprised a sloping glass roof built out from the wall, with vertical glass slides which moved on runners at top and bottom. These, and many of the other ranges mentioned in the 1861 catalogues, still survive (Figure 50).

The spatial organisation of the gardens at Somerleyton was typical. The main ranges of glass were erected, for obvious reasons, against the south-facing walls; those on the south side of the south wall – that is, outside the enclosure, and usually within the pleasure grounds – were for the display of exotic flowers. The north sides of walls – which were never lit by the sun's rays – were of little use for growing anything, and it was here that the tool-sheds and accommodation for the undergardeners were often located. On the north side of the Somerleyton gardens there were, typically, a mushroom house, a gardener's office, an undergardener's living room and a potting shed.

Kitchen gardens never fail to surprise with the ingenuity, expense and effort made to produce particular fruit or vegetables. At Great Saxham, for example, part of the north-western section of the garden was terraced and used for growing asparagus: drainage of this heavy land was improved by the insertion

FIGURE 50. The 'Conservative Wall', Somerleyton Hall.

COURTESY ANTHEA TAIGEL

of thousands of glass bottles, buried with their neck downwards, an ingenious device which now – with most still in place, but many shattered – rather limits the horticultural use of the area.

Kitchen gardens continued to be maintained, altered and embellished right through to the end of the period covered by this book. At Boulge Hall in the 1890s new slips were added to the garden, a new gardener's cottage was built, and a new door was made in the garden wall. In 1895 a new fruit and potato store was built, and a new potting shed, complete with earth closet, was constructed.[61] A continuing fascination with kitchen grounds in the 'Arts and Crafts' period was in part related to a wider concern with healthy outdoor living and rural life. This ensured a continuing demand for homegrown fruit and vegetables, even when other sources were cheap and plentiful. But even in earlier periods, domestic production was a source of pride as much as a necessity; indeed, production often outstripped consumption, as Marilyn Williams has argued, especially on estates where the family had another house 'in town'. The surplus produce was often sold off; and at Livermere in 1899, for example, it was calculated that 'sales of fruit and vegetables not required for use' raised £50 per annum.[62]

Kitchen gardens and orchards formed the main arenas of domestic production on country estates in the eighteenth and nineteenth centuries. But other facilities were closely associated with them, and often under the control of the gardener. Ice houses were a common feature, although comparatively few survive (and when they do, generally in ruinous condition). They were often, although by no means always, positioned close to a lake or other area of water. That near the canals at High House, Campsea Ash, was said in 1883 to be capable of holding '200 tons of ice'.[63] The oldest surviving example in Suffolk is probably that in Heveningham Park, its stepped gables suggesting that it may have been constructed in the late seventeenth century. Dovecotes were also still sporadically erected, probably as a conscious archaism. That at Coldham Hall, Stanningfield, for example, designed in an ornate Italianate style, lies some 140 metres to the south of the hall, proudly displayed in the open parkland.[64] It was probably erected shortly before 1838. The decorative gothic dovecote beside Cockfield Hall was built around the same time.[65] Other traditional aspects of domestic production continued in the new, leisured landscapes of the late eighteenth and nineteenth centuries. Ornamental lakes were still usually stocked with fish – some, like those at Tattingstone, are actually labelled 'fish pond' on the late nineteenth-century Ordnance Survey 6″ maps – while a number of country houses maintained fish stews, as at Langham Hall in 1842, where there was said to be 'excellent fishing, with Fish Stews for the feeding and preserving of Fish.'[66]

Suffolk's kitchen gardens have not fared well during the last half-century. Glasshouses need constant maintenance; and if they don't receive it, they fall down. Only occasionally, as recently at Thornham, have important restoration and refurbishment schemes been carried out. Only a handful of walled gardens are still under cultivation: most are now empty spaces, or are occupied by pheasant pens or Christmas trees. Some are simply choked with brambles and other adventitious vegetation. Among the most notable exceptions is the wonderful garden at Glemham House, Great Glemham, still maintained in beautiful condition, with its gravel paths lined with clipped box, a central dipping pond and nineteenth-century glasshouses ranged along the south side of the curving north wall – one, to judge from its narrow glazing bars, perhaps dating from the 1830s or 1840s. As if this wasn't enough, on the north side of the wall bothies survive, complete with fireplaces. When last visited, one still contained a bedstead.

CHAPTER 8

The late nineteenth century and beyond

New fashions, new times

Most of the parks and gardens discussed in the previous chapters were created around substantial country houses attached to large estates. True, many of their owners had made their fortunes in finance, industry or government service: Morton Peto at Somerleyton, for example, or Frank Crossley (of Crossley's carpets) who purchased the estate from him in 1861. But the creation and maintenance of extensive and elaborate grounds depended largely on rental incomes and thus, in the final analysis, on a buoyant agricultural economy. During the middle decades of the nineteenth century farming did well on the whole. In the late 1870s, however, a serious depression set in, caused mainly by competition from imports brought in from the New World – a depression which was particularly serious in an arable farming area like Suffolk.[1] At the same time, changes in political structures in Britain at both local and national level – the gradual extension of suffrage and the development of local government – ensured that the landed rich no longer dominated local life to quite the extent that they had once done. As a consequence of these and other developments, large houses and vast landscapes of fashionable display gradually lost their appeal during the late nineteenth and early twentieth centuries.

New country houses were still sporadically constructed or extensively remodelled and some new parks laid out, but where this occurred it was generally a sign that the owners had access to a large income from sources other than agricultural rents, or were themselves new recruits to the landed class from the world of commerce or industry. By the end of the century, however, most successful businessmen, while they might desire a country residence, were less keen on acquiring an extensive landed estate, with all the attendant problems and financial responsibilities. A house in the country, rather than a country house, was what many now wanted: preferably in a healthy coastal location. Houses of moderate size, built in Arts and Crafts style using mixtures of local and modern materials, and with gardens to match; or some Tudor or Stuart manor house small but reeking with rural antiquity; was what the fashionable now demanded. These were the social, economic and aesthetic forces which brought the great phase of country house landscapes to an end and ushered in a new age of garden design, largely outside the period covered by this volume.

In 1870 William Robinson published the first edition of his book *The Wild Garden*, in which he argued that hardy plants, either native or naturalised (such as bamboo), should be given more prominence in the garden, and that less reliance should be made on the kinds of half-hardy bedding-out plants which had dominated the gardens of High Victorian England.[2] Indeed, Robinson rebelled against the fashion for formal beds and argued that flowering plants should be widely scattered across the lawns and through the shrubberies and woodland belts, ideas which he developed still further in his *English Flower Garden* of 1883.[3] The most influential voice in garden design in the late nineteenth century was, however, Gertrude Jekyll, who worked with the architect Edward Lutyens from 1890, and whose first book, *Wood and Garden*, was published in 1899, followed by *Home and Garden* the following year.[4] It is on the ideas of Jekyll and Lutyens that the character of the Arts and Crafts garden largely rests.

The hallmark of 'Arts and Crafts' garden design was the combination of strong architectural elements with profuse and informal planting – mainly of hardy species and particularly in wide herbaceous borders. The 'hard land-scaping' was designed in the same vaguely vernacular style and used the same vaguely vernacular materials, as the Arts and Crafts houses created by such architects as E. S. Prior, C. F. Voysey and Lutyens himself. Steps, terraces, walls, areas of paving, pergolas, and summer houses in 'rustic' style, were all common features. These more architectural areas were laid out close to the house. Further away there was usually a series of garden compartments defined by hedges (generally of yew or privet). Some of these might be given over to various sports – there was an increasing interest at this time in healthy outdoor pursuits, and tennis, bowls, cricket and croquet were particularly popular (as was an active involvement in gardening, following the lead set by Jekyll herself). Other compartments might contain fruit gardens or collections of roses – interest in these likewise increased steadily in the last decades of the century. In the more distant areas of the grounds woodland gardens, now planted in a more naturalistic way than those of the High Victorian period, were popular. These various elements were combined to produce a timeless, rural, and quintessentially English form of garden which – in marked contrast to the more formal and garish creations previously in vogue – most people still find peculiarly appealing. They were gardens designed for relaxed rural living, appropriate to the *nouveau riches* who increasingly sought their place in the country.

Suffolk, however, was not Surrey. Far from the metropolis, it was not a place in which large numbers of Arts and Crafts houses were ever erected, not least because the appetite for a 'house in the country' could here often be supplied by the acquisition and restoration of some run-down sixteenth- or seventeenth-century manor house. Suffolk does nevertheless have some examples of houses and gardens in the new style, and a far greater number of residences more loosely influenced by it. Perhaps the most impressive example is the Lodge at Felixstowe, designed (together with the house itself)

by the renowned architect Robert Schulz Weir as a holiday home for the brewer Felix T. Cobbold.[5] It survives, but in altered form. The gardens are laid out on a series of terraces, carved out of the cliff. The most impressive feature was a buttressed retaining wall, on which was set a pergola of brick posts and oak beams; this formed one side of a sunken flower garden. Other features included a decorative dipping well, arched gateway, and wide paths of brick and pebble (Figure 51). Around the turn of the century a number of similar houses were erected by wealthy businessmen along the Suffolk coast as holiday or retirement homes. Walter Larkin, county magistrate and medical practitioner with offices in Harley Street, built a seaside retreat at Holm View, Lowestoft, which has since been demolished. The gardens laid out with his gardener, Mr MacMath, were mentioned twice in *The Gardener's Chronicle*. The journal singled out for particular praise the use of pampas grass, planted in massed clumps in long borders edged by a variety of herbaceous perennials.[6] Other concentrations of prosperous middle-class houses, displaying in their architecture and grounds some aspects of Arts and Crafts style, appeared on the outskirts of Bury St Edmunds and, in particular, Newmarket, the latter often the homes or second homes of men involved in horseracing.[7] Several

FIGURE 51. The gardens at Felixstowe Lodge, designed by Robert Schulz Weir for Felix T. Cobbold, pictured in the early years of the twentieth century.

158

were created by the developer John Watts in the last years of the century, like 'Ye Cottage', on its 4-acre plot to the east of the town, advertised for sale in 1903 with 'tennis lawn and flower border, summer house and pergola'.[8] Most examples of thoroughgoing Arts and Crafts houses and gardens, however, were created in the course of the twentieth century, and thus lie outside the period covered by this book. Moreover, the style was best suited to medium-sized residences rather than to the grounds of substantial country houses, and an exploration of such gardens thus takes us beyond our main theme. Nevertheless, as Elise Perciful has shown in her recent study of gardens of this period in East Anglia, some of the new ideas were also taken up at the very end of the century in the gardens of long-established country estates.[9]

Country house gardens

Gardens based on extensive, formal displays of bedding plants were ruinously expensive to maintain; the kinds of planting advocated by William Robinson and Gertrude Jekyll were, at least to an extent, cheaper, although the difference was perhaps rather less than is sometimes imagined. Certainly, it is not always clear how far changes in the design of particular country house gardens at this time were directly related to the need to make economies, and how far to an enthusiastic adoption of the new fashions. At Shrubland Hall, for example, an article in *The Gardener's Chronicle* in 1888 described how the gardens had been 'shorn of much of their former magnificence', and around this time many of the outlying areas of the vast grounds seem to have been grassed over, simplified or abandoned altogether. Nevertheless, alterations continued to be made, and the same article noted recent additions to the planting. 'One of the finest modern developments at Shrubland is the planting both sides of a ravine with hardy Bamboos intermixed with Dracaenas, Cannas, Maize, Funkias, and other bold and graceful plants.'[10] This was something which the 1890 edition of the *Chronicle* ascribed to the head gardener, Mr Blair, 'at the suggestion of Mr Robinson'.[11] Blair was asked 'how the plants had succeeded, and what were the kinds to be recommended for their hardihood'. He replied:

> The bamboo which you saw at Shrubland are still thriving, and have not been disturbed since they were planted two years ago. Up to this time they have grown finely, many of them forming quite a thicket. A few suffer in the foliage about the beginning of March from the cutting winds, and it is near midsummer before they quite regain their fresh green colour.[12]

He listed the particular properties of a number of varieties, including *B. Metake* ('the common sort'), *B. aure, B. viridi-glaucescans, B. mitis, B. Henomis, B. violascens* and *B. nigra*, although he noted that 'a good number of the plants were sent from Guernsey with names, but they are so much alike it would require an expert to tell the difference.' Some varieties were already as much as 20 feet (*c.* 3 metres) tall.

How much else in the Shrubland garden was changed in accordance with Robinson's suggestions is unclear: the estate archives are rather patchy for the 1880s and 1890s, and little in the way of correspondence has survived. Certainly by 1891 the pattern of planting in the Balcony Garden had been drastically altered. Drawings made for this area, showing the pattern of planting in the years 1891–94, are possibly in Robinson's own hand.[13] The *parterres* of sand, boxwork and bedding plants had been replaced by more informal beds and borders (Figure 52). These featured roses, which remained for several years, planted around with more transient annuals, including carnations, and with smaller areas of campanulas, pansies, yuccas, chrysanthemum, asters, antirinum and *gladioli anemone*. The planting seems to have become less varied and complex, and more dominated by roses and carnations, during the course of the 1890s.[14]

In the Balcony Garden at Shrubland the main framework of the old formal beds was removed when the new planting schemes were adopted. Elsewhere, earlier patterns of beds were sometimes maintained, while new forms of

FIGURE 52. Plan for the planting in the Upper or Balcony Garden, possibly by William Robinson, 1891.

COURTESY LORD AND LADY DE SAUMAREZ

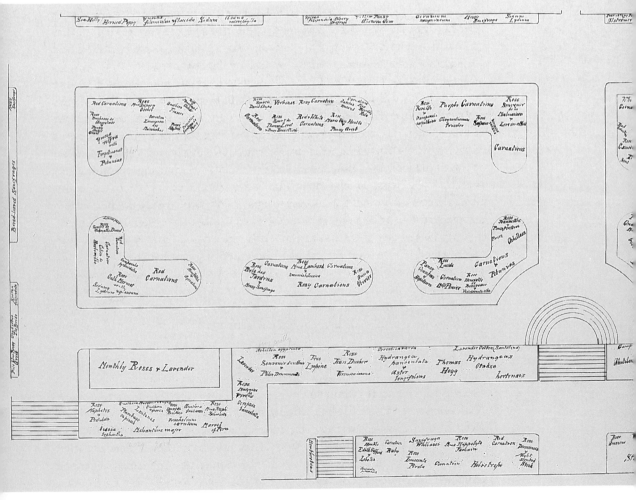

160

planting were introduced into them. At Hardwick, the famous gardens designed by D. T. Fish were thus altered, but without doing violence to the original structure. The 'very large beds cut out of turf' still survived, but they were no longer filled with bedding plants in complex patterns: photographs taken in *c.* 1900 show a variety of half-hardy plants combined with herbaceous plants and roses.[15] Something similar happened at Hengrave Hall soon after its purchase by Sir John Woods in 1897.[16] The pattern of flower beds and paths in the area to the west of the hall, originally laid out by James Howe (above, p. 134), had deteriorated in the previous decade, when the hall had been tenanted, and a photograph taken in 1897 suggests that the beds here had been grassed over.[17] Woods and his wife had the pattern cut into the lawn once again, but the beds were planted 'using a mixture of bedding and roses in a display of modern taste'.[18] The existing rose walk was altered, with the addition of many new rambler varieties.

More enthusiastic adoption of the Arts and Crafts style is apparent elsewhere. Edwin Johnstone at Rougham worked closely with his head gardener, Mr Henley, to alter and extend the gardens in the modern style. Typically, their design featured prominent architectural elements near the house (simple flower beds cut into lawns, gravel terrace etc.), and further away a series of compartments including a rose garden, flower garden, walled kitchen garden and woodland walks.[19] More impressively fashionable, perhaps, were the gardens at Boulge Hall, transformed in the 1890s soon after the purchase of the estate by the White family. In 1893 the architect Gambier Parry was paid the immense sum of £1,299 for 'Garden and terraces'.[20] Formal beds were removed: the terrace looked out over an area of lawns and specimen trees, sweeping down to a newly-constructed lake.[21] Family diaries refer in 1896 to a Dutch garden, an arborvitae walk a 'flower-bed lawn', a poplar border and a 'wilderness', presumably a woodland garden. There was a nut walk with rolled paths, rhododendron beds and a summer house. The same year two new tennis courts were constructed, and a new herbaceous border with box edging begun.[22] Equally fashionable were the grounds laid out around Cavenham Hall when this was rebuilt to designs by the Arts and Crafts architect A. N. Prentice between 1898 and 1899. They were designed by the London-based landscape gardener H. E. Milner and featured an ornamental terrace with lawn and flower beds near the house, together with yew-hedged rose garden; seven acres of walled flower, fruit and kitchen garden; a fern dell; areas of shrubbery; and extensive lawns with level areas for tennis and croquet.[23]

Although the new forms of design and planting were indeed adopted on many country estates, we should not, however, exaggerate the extent to which this occurred. Many landowners, right up until 1900 and even beyond, maintained versions of the old bedding-out schemes, albeit often on a reduced scale. Even at Shrubland Hall, while the planting of the Balcony Garden was altered, the *parterres* in the Panel Garden seem to have been maintained (they still survive, in simplified form), while at Somerleyton Nesfield's *parterres* remained in existence up to the time of the First World War. Even where country houses were entirely rebuilt and new gardens laid out in this period, many elements of the High Victorian garden were often used. Haughley Plashwood was rebuilt on a new site, 100 metres to the south-west of the old, in the 1890s, and new gardens laid out around it. These included, according to the sales particulars of 1901, an 'embroidered *parterre*' of gravel and box.[24] When Lord Cadogan rebuilt and massively extended Culford Hall around 1900 he replaced the rather informal pleasure grounds to the south of the hall, separated from the park by a ha ha and low wire fence, with a much more architectural setting, comprising a substantial Italianate terrace overlooking the park. A similar formal terrace was created to the south of the new extension built by Lord Iveagh, of the brewing family Guinness, at Elveden.[25] These were grand houses, and the homely English vernacular of the Arts and Crafts garden would not have sat easily beside them.

In spite of the increasing need to make economies, both the maintenance of extensive gardens, and in some cases the enthusiastic incorporation of elements of the new styles, were encouraged by the current vogue for gardening. Gardening was probably a more socially acceptable pastime for the wealthy, and especially for men, than it had ever been. In a report from the Bury St Edmunds Chrysanthemum Show in November 1901 *The Gardener's Chronicle* typically described how 'Some of the best known growers in East Anglia contributed plants and blooms, the successful exhibitors including Sir S. B. Crossley, Bart. MP; Lord de Ramsey; Sir James Miller, Bart.; Sir J. Blundell Maple MP; Mr W. R. Seago of Oulton Hall Lowestoft; the Marquess of Bristol and the Hon W. Lowther.[26] A number of Suffolk landowners were particularly noted for their horticultural passions. Colonel Rawson, for example, avidly collected orchids and ferns at Coldham, steadily expanding the ranges of glass in the gardens there. Colonel Powell at Drinkstone was similarly noted for his orchids and roses; while Lord Cadogan and his wife maintained a nationally important collection of zonal pelargoniums at Culford.[27]

The late nineteenth century thus saw a variety of developments in the gardens of Suffolk's country houses. There was less change in the wider landscape of the park. Although the agricultural recession may have encouraged the simplification and grassing-down of expensive gardens, it gave little incentive to plough up parklands. The main change, in fact, was a marked expansion in the area of woodland, principally effecting parks located on the sandier soils of the county, especially in Breckland. Landowners in this arid region had long been more involved than most in game-shooting; many now exploited the commercial possibilities of this pastime on an increasing scale, leasing their estates for periods to wealthy individuals from the metropolis, and devoting large areas to game-rearing rather than agriculture. New areas of woodland, designed principally as pheasant cover, consequently encroached upon formerly open areas of parkland. The numbers of game taken by the end of the century in and around Breckland parks were phenomenal. Livermere, leased by the de Saumarez family to Charles Reece in 1899, was said to have had an average 'bag' of 2,678 pheasants, 635 partridge, 297 hare and 220 rabbits. At Elveden, 70 workers were employed as gamekeepers or their assistants, rearing 20,000 pheasants each year.[28]

Old manor houses with new grounds

As already noted, one of the most distinctive features of the late nineteenth century was the tendency for individuals who had made their money in commerce, industry or the professions to purchase not a large country house and its attached estate but some small Tudor or Stuart manor house which had long since declined in status to a small working farm. This they would restore, adapting surviving walled gardens in the modern style and sometimes laying out a small park at the expense of the surrounding fields. Suffolk had a number of ruinous, picturesque manor houses of this kind, and with agriculture in a state of recession several were up for sale at a reasonable price. The Leicester architect Arthur Wakerley thus bought Gedding Hall Farm in 1897. He carefully restored the building and laid out a suitably compartmentalised garden around it. Particulars drawn up when the family sold the estate in 1918 describe gardens which afforded 'endless scope for a keen horticulturalist', although owing to the recent war they were not kept up as in former times. There was a 'tennis lawn, partly walled vegetable and fruit garden, double glasshouse, rose and flower garden, orchard and spacious thatched summer house'. Part of the moat had been drained and was used for recreation, initially as a bowling green, later as a rifle range. Several of the pasture fields around had been thrown together to make a small park – little more than an extensive pasture containing former hedgerow trees.[29]

More dramatic was the transformation of Stutton Hall, overlooking the estuary of the river Orwell in the far south of the county. This sixteenth-century house and its garden have already been discussed (p. 13). By the nineteenth century the property had declined in status, and in 1856 'Stutton Hall Farm' was put on the market as a 'most substantially built family residence' with a good range of farm buildings.[30] It was back on the market in 1887, and was purchased by James Oliver Fison, the agricultural chemical manufacturer. He restored and extended the house, repaired the old walled garden and laid out fine new gardens in a vaguely Arts and Crafts style to the south of the house – an area of lawn with lines of topiaried yews (some Irish) and a number of distinct compartments defined by neat yew hedges.

Particularly striking, however, was the new parkland which Fison created around the hall. The First Edition Ordnance Survey map of 1884 shows a landscape of fields and farmland; by 1905, a landscape park, covering some 45 hectares, had appeared. This was approached from the north by a long, straight drive, which perpetuated the line of an earlier avenue. Fison had the edges of this feature planted up with mixed belts containing Scots pine, horse chestnut, sweet chestnut and sycamore, together with some Corsican pine, oak and beech. The park, in the traditional manner, contained a number of old oaks retained from the earlier hedgerows but was mainly planted up in lively fashion with a mixture of oak, horse chestnut, sweet chestnut, Wellingtonia and (most striking of all) cedars – Cedar of Lebanon, Deodar cedar and a number of examples of *Atlas glauca*, paired in several places with horse chestnut – an unusual but effective combination. Belts still run along the northern edge of the parkland planted with Corsican pine, Scots pine, chestnut and oak. The result is original and striking, the last gasp of the landscape park tradition.

There are a number of similar examples. Thus the picturesque early Tudor manor house of Giffords Hall near Stoke by Nayland, with its walled gardens of seventeenth-century date, was bought by J. W. Brittain in 1888. At this date the hall was surrounded by a landscape of hedged fields, but by 1903, according to the second edition OS 6″ map, a sparsely-timbered park of *c.* 100 acres (*c.* 40 hectares) had been created. At Nether Hall, Pakenham, the 'park-like paddock' of 10 hectares shown on the sales map of 1865 (above, p. 00) [31] had been doubled in size, its farm buildings swept away, and a lodge built at the north entrance by the time of the First Edition Ordnance Survey in 1883.

Cuthbert Quilter and Bawdsey Hall

Old manor houses might be upgraded in this period; the residences attached to existing landed estates might occasionally, as at Culford, be rebuilt. But only very rarely were entirely new country houses erected on virgin sites. The most striking example in Suffolk is unquestionably Bawdsey Manor, which occupies an exposed and isolated coastal location beside the North Sea, some 12 km south-east of Woodbridge. Strictly speaking, the full development of this site takes us slightly outside the chronological limits of this book, but this is a temptation impossible to resist.

In spite of its imposing name, Bawdsey Manor does not occupy the historic site of the manor of Bawdsey which, to judge from an estate map of 1727, lay within Bawdsey village, beside the parish church.[32] Indeed, the 1840 Tithe Award map shows its site quite empty of buildings, except for a martello tower (a coastal defence of Napoleonic age), although a farm and a cottage lay at no great distance.[33] The house was the creation of Sir Cuthbert Quilter and began life in the 1880s as a small holiday home conveniently located at no great distance from the family's principal residence at Hintlesham Hall to the west of Ipswich.

Quilter was born in 1841, and at the age of seventeen followed his father into the accountancy business. At twenty-two he became a stockbroker and later created the National Telephone Company, which was eventually taken over by the GPO to Quilter's great financial advantage. In the 1890s he decided to make Bawdsey the main family residence, and the house was extended piecemeal in a number of stages. The Red Tower was added in 1895, the White Tower was completed in 1904, and the two were linked in 1899. The result is a curious building in a strange mixture of styles – gothic, Elizabethan, and Jacobean. At the same time, Quilter was busy acquiring land in the area around the house, eventually building up a substantial estate (reputedly a total of over 3,000 hectares). He also acquired the title of Lord of the Manor of Bawdsey – hence presumably the name of the house. Quilter entered fully into the spirit of country life, constructing new farm buildings and a number of new farmworkers' cottages. From 1885 until 1906 he was MP for Sudbury.[34]

The elaborate gardens laid out around the new house were largely the creation of Lady Quilter, although Sir Cuthbert also appears to have been actively involved. They were in a medley of styles, but their somewhat idiosyncratic character was largely determined by the peculiar qualities of the site. In the words of the *The Gardener's Chronicle* for December 1908:

> It would be difficult to find a more windswept spot even on the east coast than that which was selected for this residence ... Modern engineers appear capable of building handsome houses under almost any conditions, and they can make their base what they wish it to be, but it is another matter altogether when attempts are made to carry out first-class gardening under conditions absolutely opposed to the growth of all but the very hardiest species of plant.

The local winds, the article went on to assert, were 'often sufficient to blow a plant out of the earth altogether'. Shelter was thus of supreme importance in the garden's design; so too was an emphasis on the more hardy alpine plants.[35]

Immediately below the house, on its south-western side, was (and is) a series of vaguely Italianate terraces in red brick, leading down to a level grassed area which is used now, as in Quilter's day, as a cricket pitch. A substantial boathouse is built into its base, and on its upper tier there is an elaborate tea house, in vaguely Jacobean style. Its walls were originally lined with decorated tiles collected by Sir Cuthbert from southern Italy, although many have since been removed. The terraces and associated structures survive in good condition, although the area is now mostly laid to grass.

On the north-eastern side of the house are the remains of a circular sunken garden, originally a rose garden. From here, according to the article in *The Gardener's Chronicle*, the sea could always be heard but never seen. This garden was formed by Sir Quilter 'on the exact site of an old coastguard station [i.e. the martello tower], which he first had to blow up with explosives' – clearly, this was an age before English Heritage, Listed Buildings regulations and the rest! As well as roses the garden contained areas of bedding plants, hydrangeas in tubs and carnations in stone vases. A circle of yew trees flanked the central area, four of which still survive. The garden was linked to other parts of the grounds by grotto-like underground tunnels, partly constructed with materials from the demolished tower. These are still in existence.

One of these tunnels still leads to the most striking feature of the gardens: the 'rockery wall' or artificial cliff, 50 metres high and extending along the shore for some 400 metres. It is a massive feature, constructed of cement generously embedded with artificial 'Pulhamite' stone and local crag (Figure 53). It was

FIGURE 54. Bawdsey Manor: the pergola garden.

intended to provide both a habitat for alpine plants, and to some extent shelter for the garden, for at its summit it was raised, as a low bank, a metre or more above the level of the land. The original planting was, according to *The Gardener's Chronicle*, based on 'plants most likely to thrive under the unusual conditions to which the rockery is exposed, rather than … risking a number of choice Alpine species with little hope of their ever getting satisfactorily established'. The article noted species of *thymus, iberis, centranthus, gypsophila, Lupinus arboreus, eryngium, hydrangea, calendula, aubretia*, and *Fuchsia Riccartonii*, among others.[36] A precipitous path still leads along the side of the cliff several metres below its summit, threading in and out of alcoves and cave-like recesses. On the whole the rockery cliff is in good condition, although it is now very overgrown and has suffered (somewhat picturesque!) collapse in a few places.

Along the top of the 'cliff' a terraced walk was laid out, between the counterscarp bank and the wall of the extensive kitchen garden. This was originally lined with yews, some of which survive. It led to the Pergola Garden (Figure 54), lying immediately to the east of the sunken rose garden. This featured a substantial pergola and a lily pond. The former, with stone pillars and a wooden top, was described in 1903 as 'almost completely furnished with such plants as Honeysuckle, Clematis, Ficus, species of Rubus, Periploca, and

FIGURE 53 (*opposite*). Bawdsey Manor: the artificial cliff.

other suitable genera'. This garden is in a less happy state than most of the site: some of the original trees survive but the pond is dry, the pergola has lost its roof, and some of its pillars are leaning badly, but the paths and paving survive largely intact.

Beyond this area, to the west (and thus further away from the sea), the Quilters constructed a large kitchen garden covering around 2 acres, complete with an orangery of wood and glass. The walls survive in good condition, but the interior has now been laid to grass, and the fine, classically detailed orangery is badly decayed and approaching the end of its life. Beyond the gardens proper there were parkland grounds of *c.* 60 hectares, occupied by grass lawns and plantations, featuring in particular pines, holm oak, sweet chestnut and sycamore.

Bawdsey Manor and its grounds were taken over by the RAF in 1936 and used as a research station where radar was developed. After 1939 the site was used as a radar station; subsequently it became an RAF training school, a function it performed until 1974. It was then closed for four years, reopening in 1979 as an air defence surface-to-air Bloodhound missile unit. The base was finally decommissioned in 1991. Throughout this period house and immediate grounds were maintained in remarkably good condition, but the surrounding parkland is now largely occupied by the usual collection of military detritus: buildings, roads, block houses etc. The 'cliff' has a number of gun emplacements built into it, making for a particularly striking and complex piece of twentieth-century archaeology.

Bawdsey, with extensive grounds, was the most impressive new country house to be erected in Suffolk in this period. No other site really compares with it. Although in one or two other places new parks came into existence around new country houses, these were on a very different scale. Bentley Park, for example, did not exist when the First Edition OS 6″ map was surveyed in 1884; a small farmhouse or cottage occupied its site, and the land around was divided into arable fields. By 1905, however, 'Bentley Lodge' had been built, a medium-sized Edwardian house of brick with a slate roof, surrounded by a simple park of *c.* 30 hectares, little more than an area of grass paddocks scattered with trees and a few modest clumps.

Public parks and gardens

As the agricultural depression deepened, few country houses were built or rebuilt and few attempts were made at large-scale, flamboyant landscape design. The great age of country house gardening was coming to an end. Yet, by a strange twist of fate, aspects of these landscapes of privilege were given a new lease of life in the parks and pleasure grounds laid out for a very different clientele – the general public.

To a large extent, the provision of parks grew out of the wider Victorian interest in reform and social improvement.[37] Urban growth and the enclosure of commons and wastes had removed much of the population from the open spaces in which they might enjoy healthy exercise; parks were commonly referred to as the 'lungs' of a town.[38] Genuine concern for an urban working class afflicted by poverty, squalor and poor sanitation was combined on the part of decision-makers with the fear of the potential barbarism lurking beneath the surface of a population brutalised by industrialisation and urbanisation. Parks were arenas where the urban population could be educated and socialised. Of course, motives other than philanthropy or social control also operated. Many urban parks were part of middle-class housing developments and were a way of making these more appealing to potential customers. In effect, they provided residents with some of the trappings of a landscape park, and this is one of the reasons why many of the principal features of urban parks were derived from the pleasure grounds of the landed rich.

The Report of the Select Committee on Public Walks emphasised the physical need for open spaces and the problem of working-class recreation in 1833, and the Municipal Corporations Act of 1835, which allowed local authorities to levy a rate, led to the creation of a number of early parks, such as Moor Park in Preston, Lancashire. But many early parks, such as Derby Arboretum (1840), were not provided by local authorities (or created by developers) but were the gift of local philanthropists. Even in the middle and later decades of the century private philanthropy paid for many parks, although local authorities increasingly acquired land for public recreation, sometimes by raising money by subscription so that the purchase price was not a charge of the municipality or the rate-payers. It was only with the passing of the Recreation Grounds Act of 1859, which encouraged the donation to the local authorities of land (or money) for recreation; and of the Public Improvements Act of 1860, which allowed authorities to acquire, hold and manage land for purposes of recreation; that urban parks really began to appear on any scale. There was much debate throughout the nineteenth century about the precise purpose and layouts of public parks, but most designs placed an emphasis on horticulture, perambulation and quiet 'improving' recreation rather than on sports.

Belle Vue Park at Lowestoft was in many ways typical of the parks laid out in late Victorian times. It occupies a level central area which slopes precipitously towards the east – in the direction of the North Denes and the sea – and towards the north, where a deep natural ravine called Gallow's Score is now followed by a public road. In the late eighteenth century, part of this area had been occupied by a gun battery, dominating the Denes and the anchorage beyond. Most, however, was common land, part of North End Common, and in the late nineteenth century was used, among other things, as a communal drying ground.[39]

Lowestoft was expanding fast in the later nineteenth century, both as a holiday resort and as a fishing port and harbour. At a meeting of the Improvement Commissioners in 1872 reference was made to the North End Common having become 'the resort of rough and disreputable characters'. The then chairman of the Commission, James Peto, expressed a hope that 'the time was not too distant when the town would act wisely and make the common into an ornamental recreation ground' – not least because the area lay close to an expanding area of middle-class housing. The Commission took his advice, and two members – William Youngman and the solicitor William Rix Seago – worked hard to realise this aim. The park was opened in 1874 at a ceremony performed by the lord of the manor, Mr Robert Reeve, and his wife. Typically, its layout mirrored that of contemporary pleasure grounds of the wealthy, with paths winding through areas of lawn and shrubbery, and with an ornamental pagoda erected on the eastern side, commanding views across the sea. This has been removed – a substantial memorial to those who fell in the 1939–45 war now occupies its site – but the park otherwise survives in fine condition, a rather neglected piece of Suffolk's garden heritage. A lodge in *cottage ornée* style still stands at the original, south-western entrance, thatched, with elaborate chimneys, rusticated woodwork and decorated barge-boards (Figure 55). (It was substantially rebuilt after a disastrous fire in the late 1980s.) It provided a home for the park-keeper, who ensured that only the more respectable elements of the population, using the park in an appropriately sedate manner, would be admitted. The entrance path divides immediately, running either side of a small sunken garden with box hedging and with golden Irish yews at the corners. The paths then meander through

FIGURE 55. Belle Vue Park: the lodge.

lawns with island beds and areas of woodland/shrubbery (Figure 56). They are edged in places with artificial rock work. Originally, the park was planted with a variety of specimen trees, but its exposed position beside the North Sea has ensured that only the hardier varieties have survived – Scots pine, Corsican pine, Western red cedar, beech, horse chestnut and many fine holm oaks. These trees are underplanted with yew, holly, Portugal laurel, broom and rhododendron On the northern side of park, as already noted, the ground falls steeply into a ravine. A number of paths were laid out along the slope, some of which survive, but the most striking feature here was added in 1887, when a member of the town council, Mr Arthur Stebbing, suggested that the Queen's Jubilee should be celebrated by building a bridge across the ravine, thus connecting the park with the middle-class houses which had recently been built in the North Parade.

> Assisted by several other gentlemen, he raised a goodly sum. But Mr William Youngman JP, as Alderman of the Borough, who was its first elected Mayor, and who has also benefited his native town to a great extent, came forward and gave the bridge, which stands as a substantial and useful reminder of the Queen's Jubilee, and of Mr Youngman's munificence.

The bridge was designed by Richard Parkinson, who was chief engineer to the Eastern and Midlands Railway, one of the lines which later made up the Midlands and Great Northern Joint Railway. In 1897 the Lowestoft Corporation purchased an adjoining area of land, the Sparrows Nest Estate, to add to the park. It owed its name to the small holiday cottage built here by Robert Sparrow of nearby Worlingham Hall. This area has seen much twentieth-century development, and only traces remain of the nineteenth-century planting, again featuring holm oak and yew.

Suffolk has few other public parks originating before the turn of the twentieth century, a reflection of its essentially rural character. Interestingly, the earliest (with the exception of Belle Vue) began life as private parks attached to country houses. Christchurch Park in Ipswich thus began as the park and pleasure grounds of Christchurch Mansion on the northern fringes of the town. Since the eighteenth century, the grounds had been open to the 'respectable' public on a sporadic basis, but in 1894 house and grounds were donated to the town by the wealthy brewer Felix Cobbold. Adjoining land was purchased for £26,000, and the park was opened in 1895. New additions had been made to the landscape, which included ornamental flower gardens, a mount, tennis courts and bowling greens, and an arboretum. In 1897 the mayor entertained 10,500 children there as part of the town's jubilee celebrations.[40] Some of the other main parks in the town have early origins as country house landscapes engulfed by the steady outward spread of the town and acquired by the council for recreation. Holywell is a fascinating site, incorporating a string of reservoirs and a canal created in the 1740s to provide water to the new Cobbold brewery on Cliff Quay, and vestiges of nineteenth-century gardens and pleasure grounds. It only became a public park in 1935, when it was presented to the County Borough by Lord Woodbridge. Chantry Park, with its many fine specimen trees and lime avenue, again began as a private residence and only became a public park in 1927.[41]

Into the twentieth century

The agricultural depression lifted slightly during the First World War, but in its aftermath many landed estates were sold and broken up. Those on the poor, sandy soils of Breckland fared particularly badly. Much of their land was purchased by the Forestry Commission in the 1920s and 1930s; their mansions were demolished and their parks, belts and clumps engulfed in conifer plantations. Many were already in a dire state when purchased, as the reports made by the Commission before purchase make clear. The Downham Hall estate had been in the hands of the Board of Trade Timber Supply Department during the War, and had subsequently passed through the hands of a series of owners. Farms were unlet, much of the farmland derelict, and the mansion in disrepair. Parkland had fared particularly badly:

In consequence of the timber operations the park, as such, has ceased to

172

exist. An avenue of limes and a few quasi-ornamental trees of little com-mercial value are all that are left, except in some of the wild belts where a few ragged conifers remain, the best timber having been cut out.[42]

Decline continued in the inter-war years, and the complex gardens of the nineteenth century, where these still survived, were progressively simplified. During the Second World War there was further decline: many parks were ploughed as part of the 'Dig for Victory' campaign and many park woodlands felled to meet the national timber shortage. Agricultural fortunes revived in and after the war, but those of country house landscapes did not. In the political circumstances of 1945, a country seat set in a vast and useless landscape no longer inspired awe and brought political advantage; such things were an embarrassment, an encumbrance, an invitation to the taxman. Even as landed fortunes revived further during the 1950s and 1960s country houses were often wholly or partly demolished, their owners preferring to live in more modest but more convenient premises.

The great landscapes described in this book thus survive today in a variety of conditions (Figure 57). Some have almost entirely disappeared, their mansions demolished and their parks ploughed, quarried for sand and gravel, or given over to forestry. Such a fate has mainly befallen those on the light soils of Breckland (Livermere, Santon Downham, Fornham St Genevieve) and the Sandlings (Sudbourne), but a number on the heavy claylands of the north and east have fared no better (Redgrave, Flixton, Hoxne). Elsewhere the house survives (although now often in divided ownership or institutional use) but the park is under the plough and in degraded condition, as at Benhall, for example, or converted into a golf course, as at Hintlesham. In a surprising number of cases, however, the parkland remains largely intact, together with the gardens, although the latter have normally been much simplified. Notable examples include the great show-piece sites like Shrubland and Somerleyton, and more private, less well-known places like Benacre, Sotterley, Glemham Hall, Glemham House and Hengrave. Occasionally, and bizarrely, the parkland survives unploughed even though the house has long gone – as at Henham and Thornham. Only in a few places have ambitious schemes of restoration and embellishment been undertaken, most notably perhaps at Heveningham, where in the 1990s the gardens were extensively altered and the park lakes restored and extended.

What will be the future of these historic landscapes? A burgeoning interest in 'our' heritage, and the current tendency to move subsidies away from agriculture and into environmental conservation suggests that parks, at least, have a better medium- and long-term future than they have had for many years, although development pressures – and in particular the needs of the leisure industry and 'conference centre culture' – will remain a threat. Gardens are a different matter, for their upkeep is more demanding and their fabric more fragile. But public enthusiasm – channelled effectively by the Suffolk Gardens Trust – will doubtless help to ensure the survival of much of what is left.

Although there is some financial support for the conservation of historic landscapes (through such things as the Countryside Stewardship Scheme), there is very little statutory protection for them. Even the most famous parks could be ploughed up at the owner's whim, and although in certain areas tree preservation orders can protect the main planting, and buildings conservation legislation can protect aspects of the 'hard landscaping', the more ephemeral features can be allowed to decay or removed with impunity. It is a testament to the enthusiasm of owners – part of the shared twentieth-century obsession with the past – that so much survives. The only official recognition by central government of the importance of designed landscapes is the English Heritage *Register of Parks and Gardens of Special Historical Interest*. This document offers no statutory protection but must be invoked as a consideration when plans for development, road improvement and the like are being considered by the

FIGURE 57. The site of Flixton Hall, with earthworks of the gardens made by William Andrews Nesfield.

COURTESY DEREK EDWARDS AND NORFOLK MUSEUM SERVICE

appropriate bodies. The Suffolk volume, however, is a farce: it contains a mere 7 sites, and many of the finest landscapes in Suffolk – ones which have been referred to again and again in the pages of this book – do not feature in it: places of immense archaeological and historical importance, like Glemham House, Sotterley, Great Saxham and Hengrave.

It is arguable that tighter legislation and more rigorous protection is required. Certainly more Suffolk sites ought to be included on the *Register* as a matter of some urgency. But protection and financial support need to be targeted, and this requires more understanding of the character of Suffolk's garden heritage. In my own view, this involves some comprehension of what is special and distinctive about the county's garden heritage: what makes it different from that of Norfolk, Essex, Cambridgeshire or areas further afield.

This book will, I hope, have supplied some of the answers to this question. But it might be useful to rehearse here one or two of these distinctive features, all of which ultimately derive from Suffolk's particular environmental, social and economic characteristics. One might instance the local enthusiasm in the decades around 1700 for canals – more probably related to the earlier taste for moats and fish ponds, and to the heavy nature of much of the county's soils than to direct Dutch influence. For the early and middle eighteenth century, the large number of parks created as settings for houses – before the advent of Capability Brown – is noteworthy; so too is the paucity of sites which display a confident familiarity with the kinds of trendy serpentine pleasure grounds so common in Hertfordshire or other counties close to the metropolis. For the later eighteenth century, what is striking is the large number of early Repton sites, and the general proliferation of small and medium-sized parks. The two may well be connected, Repton's familiarity with the needs of the new park-makers at the start of his career being a major factor in the emergence of his distinctive style. In the nineteenth century, Suffolk held a key place in the development of fashions in planting, with the work of Beaton and Lady Middleton at Shrubland, and the county has more than its fair share of gardens designed by William Andrews Nesfield. The vitality of Victorian gardening – which contrasts, to some extent, with the situation in neighbouring Norfolk – was underpinned no doubt by a flourishing economy in all parts of the county during the High Farming period. On the other hand, the essentially rural character of the county ensured that – with a few obvious exceptions – little large-scale landscaping occurred in the period after *c.* 1880, and the county's heritage of Victorian public parks is rather meagre.

Rather different aspects of Suffolk's distinctive heritage will doubtless occur to readers. Others will emerge when further research has been carried out. For while some of the sites mentioned in this volume have been studied and analysed in considerable detail – such as Shrubland and Somerleyton – others have received a more cursory inspection and would certainly repay more detailed study. Moreover, as I have emphasised throughout, this book has been almost exclusively concerned with country house gardens; research on

Suffolk's smaller gardens is urgently needed. All in all, a vast field awaits exploration, and this volume is an introduction to, rather than the last word on, a fascinating subject. I hope that others will take this research much further. Investigating garden history is a pleasure in its own right, but it is also of vital importance in understanding the character of what survives, and thus for formulating decisions about what should be preserved.

Of course, it is a moot point whether we should care over-much about the future of these relics from a lost age of ostentatious privilege. In Suffolk the answer is perhaps more straightforward than in some other counties. In an intensively agricultural countryside, in which woods, trees and hedges have been lost at an appalling rate since the 1960s, parks often provide islands of pasture which are havens for wildlife and often a precious archaeological resource. In parks like Sotterley we can stand among oak trees which once grew in a densely-treed, wood-pasture landscape which has been largely eradicated. It is arguable, too, that as expressions of the taste of a ruling elite, gardens and parks are themselves an archaeological resource of considerable importance, through which we can read the changing character of social relations. Like other aspects of material culture, they can help us understand the workings of past societies, and their preservation need not imply any condoning of the massive social and economic inequalities they manifested and helped perpetuate. But beyond all this, where – through the enlightened actions of owners or as conditions of compliance with various government schemes – parks and gardens are open to public view, on a sporadic, frequent or regular basis, then it is possible to enjoy these landscapes in a much more basic, sensory way. To stand among the ancient sweet chestnuts or look down Barry's breath-taking Descent at Shrubland Hall, to wander through the quiet, intimate landscape of Thornham, or to watch the wildfowl wheeling above the great expanse of Repton's Henham, are immensely enjoyable experiences, tempered only by the knowledge that many equally wonderful places remain out of bounds to you and me.

Notes

Notes to Chapter 1. The context of garden design

1. W. G. Clarke, *In Breckland Wilds* (London, 1925).
2. M. R. Postgate, 'The Field Systems of Breckland', *Agricultural History Review* 10 (1962), pp. 80–101; S. Wade Martins and T. Williamson, *Roots of Change: Farming and the landscape in East Anglia, c. 1700–1870* (Agricultural History Society Monograph, Exeter, 1999), pp. 12–16.
3. ESRO Iveagh MS HD 1538/212/1.
4. Wade Martins and Williamson, *Roots of Change*, pp. 34–43.
5. Lord F. Hervey (ed.), *Suffolk in the Seventeenth Century: the Breviary of Suffolk by Robert Reyce* (London, 1902), p. 29.
6. WSRO, T1/16. For more information about the development of the Suffolk clayland landscape in the seventeenth and eighteenth centuries see Wade Martins and Williamson, *Roots of Change*, pp. 21–7, 67–9.
7. The best account of the Sandlings remains: E. D. K. Burrell, *Historical Geography of the Sandlings of Suffolk, 1600–1850* (unpublished MA Dissertation, University of London, 1960).
8. Arthur Young, *General View of the Agriculture of Suffolk* (London, 1795), p. 38.
9. *Humphry Repton: the Red Books for Brandsbury and Glemham Hall*, facsimile with an introduction by Stephen Daniels (Dumbarton Oaks: Washington, 1994).
10. Richard Blome, *Britannia* (London, 1673), p. 428.
11. Edward Martin, 'Great Houses of the Sixteenth and Seventeenth Centuries', in D. Dymond and E. Martin (eds), *An Historical Atlas of Suffolk* (Suffolk County Council and Suffolk Institute of History and Archaeology: Ipswich. Revised and enlarged edition, 1999), pp. 92–3.
12. J. Thirsk, 'Seventeenth-Century Agriculture and Social Change', in J. Thirsk (ed.), *Land, Church and People: Essays Presented to Prof. H. P. R. Finberg* (Agricultural History Review Supplement, Reading, 1970), pp. 148–77.
13. Humphry Repton, *Red Book for Culford Hall*, 1790: private collection.
14. WSRO 317/1.
15. Humphry Repton, *The Red Books for Brandsbury and Glemham Hall*.

Notes to Chapter 2. Parks and gardens before *c.* 1660

1. W. A. Coppinger, *The Manors of Suffolk* (Manchester, 1911), vol. 7, p. 200.
2. John Gage, *History and Antiquities of Hengrave* (London, 1822), p. 17.
3. 'Hengrave Hall, Suffolk', *Country Life* 27 (1910), pp. 558–68; Plan of Hengrave Park, 1588: WSRO P746/1.

4. Map of Hoxne New Park, 1619, ESRO HD40 422; Map of Cranley Hall, Eye, private collection.

5. The Lands and Grounds Belonging to Badingham Hall, 1614: WSRO P462 6392.

6. RCHME, National Monuments Record, G1174.

7. C. Richmond, *John Hopton: A Fifteenth-Century Suffolk Gentleman* (Cambridge University Press, 1981), p. 148. The stews were noted as 'newly built'.

8. Lord F. Hervey (ed.), *Suffolk in the Seventeenth Century*, p. 50.

9. C. Taylor, 'Moated Sites: their Definition, Form and Classification', in F. A. Aberg (ed.), *Medieval Moated Sites* (Council for British Archaeology: London, 1978), pp. 5–13.

10. ESRO HD 40 (422); Barrow is shown on a copy of a map of 1597, made in 1779: WSRO 862/2.

11. ESRO HD 417/8: map of Bedingfield Hall, 1729.

12. *Country Life*, December 1898; August 1956.

13. C. Taylor, P. Everson and W. Wilson-North, 'Bodiam Castle, Sussex', *Medieval Archaeology* 34 (1990), pp. 155–7; P. Everson, '"Delightfully surrounded with Woods and Ponds": Field Evidence for Medieval Gardens in England', in P. Pattison (ed.), *There by Design: Field Archaeology in Parks and Gardens* (Royal Commission on Historical Mounments: London, 1998), pp. 32–8.

14. C. Taylor, *Parks and Castles of Britain: a Landscape History from the Air* (Edinburgh University Press, 1998), pp. 40–1.

15. P. Murphy, pers. comm.

16. J. Kenworthy-Brown, P. Reid, M. Sayer, D. Watkin, *Burke's and Savills Guide to Country Houses, vol. III, East Anglia* (London, 1981), p. 264.

17. Kenworthy-Brown *et al.*, pp. 252–3; A. Oswald, 'Melford Hall', *Country Life* 82 (1937), 116–21, 142–7.

18. 'A platte and description of the Lordship of Long Melford', 1615, for Sir Thomas Savage; 1613 map by Samuel Piers; both at Melford Hall.

19. See A. Hassell Smith, 'The Gardens of Sir Nicholas and Sir Francis Bacon: an Enigma Resolved and a Mind Explored', in A. Fletcher and P. Roberts (eds), *Religion, Culture and Society in Early Modern Britain* (Cambridge University Press, 1994), pp. 125–60, 144.

20. The garden enclosure is not shown on a map of 1597, a copy of which is in the possession of the owners, but the brickwork (small, moderately regular bricks laid in an irregular English bond) and the 'Tudor' arches to the gate in the west wall and above the alcoves at the eastern end of the north and south walls suggest it was constructed not much later.

21. R. Strong, *The Renaissance Garden in England* (Thames and Hudson: London, 1979); John Dixon Hunt, *Garden and Grove: the Italian Renaissance Garden in the English Imagination 1600–1750* (Dent: London, 1986).

22. ESRO 295; ESRO 942.64 Som.

23. J. Freeman (ed.), *Thomas Fuller: the Worthies of England* (George Allen and Unwin: London, 1952), p. 523.

24. My thanks to A. Hassel Smith for providing me with this quotation: Raynham Hall (Norfolk) Attic: 'Accounts 1633–34 manorial, estate and misc'.

25. The literature on medieval parks is extensive. See, in particular, O. Rackham,

History of the Countryside (Dent: London, 1986), pp. 122–9; P. Stamper, 'Woods and Parks', in G. Astill and A. Grant (eds), *The Countryside of Medieval England* (Blackwells: Oxford), pp. 124–48; J. Birrell, 'Deer and Deer Farming in Medieval England', *Agricultural History Review* 40, 2 (1993), pp. 112–26.

26. R. Hoppitt, *A Study of the Development of Parks in Suffolk from the Early Eleventh to the Seventeenth Century* (unpublished PhD Thesis, University of East Anglia, 1992).

27. *Ibid.*, p. 82.

28. *Ibid.*, p. 71.

29. *Ibid.* pp. 63–5.

30. *Ibid.*, p. 85.

31. D. Dymond and R. Virgoe, 'The Reduced Population and Wealth of Early Fifteenth-Century Suffolk', *Proceedings of the Suffolk Institute of Archaeology and History* 36 (1986), pp. 73–100.

32. Hoppit, *Parks*, p. 89.

33. *Ibid.*; see also G. F. Peterken, 'The Development of Vegetation in Staverton Park', *Field Studies* 2 (1968), pp. 1–39; Norden's Survey of the Manors held by Sir Michael Stanhope, ESRO I V5/22/1.

34. Hoppitt, *Parks*, pp. 129–35; PRO MPC1, MPC2, MPC3.

35. ESRO HD 417/14.

36. *Ibid.*, p. 91.

37. Hassell Smith, 'The Gardens of Sir Nicholas and Sir Francis Bacon', p. 144.

38. WSRO 449/3/15. Earlier maps (1742: WSRO 712/58, and 1769: WSRO 712/60) show that the avenue widened into a circle at this point.

39. Hoppit, *Parks*, pp. 253–4.

40. Map of Hoxne New Park, 1619, ESRO HD40 422.

41. Hoppit, *Parks*, p. 70.

42. Map of Melford Hall and Park, 1580, by Israel Amyss, at Melford Hall.

43. 10 Jac. 13 Pars Rot 167.

44. Hoppitt, *Parks*, vol. 2, p. 17.

45. 1613 map by Samuel Piers, at Melford Hall.

46. J. Cullum, *History and Antiquities of Hawsted in the County of Suffolk* (second edition, with additions by T. Cullum, London, 1813); Hoppitt, *Parks*, pp. 269–70; A Surveye of the Mannor of Hawstead with Buckenhams, *c.* 1616: WSRO E8/1/3.

47. Plan of Hengrave Park, 1588: WSRO P746/1.

48. It has been replanted at least twice. Most of the constituent trees are nineteenth-century limes, but at the far, southern end, where eighteenth-century maps show that it widened into a circle (probably surrounding a viewing mount), limes with girths of 5 metres suggest a replanting in the early eighteenth century.

49. WSRO HA 528/31.

50. Made for an uncertain purpose: WSRO Hengrave manorial 449/5/31/36.

51. A description of the county made by an unknown observer in *c.* 1600: D. Mac-Culloch (ed.), *The Chorography of Suffolk* (Suffolk Record Society/Boydell: Ipswich, 1976).

52. Helmingham Hall archives, B1/11/1; B4(a) 18; B1/28/26.

53. ESRO 295; ESRO 942.64 Som.

54. A Surveye of the Mannor of Hawstead with Buckenhams, *c.* 1616: WSRO E8/1/3.
55. WSRO E8/1/3.
56. I am grateful to Catherine Bond for information about the Kentwell ponds.
57. W. A. Coppinger, *The Manors of Suffolk* (London, 1905), vol. 1, p. 143.
58. Sir William Hyde Parker, *Melford Hall and Manor* (London, 1873), p. 332.
59. *Private Correspondence of Jane Lady Cornwallis 1613–1644* (Bentley, Wilson and Flay: London, 1842), p. 57.
60. *Ibid.*, p. 164.
61. J. Gage, *History and Antiquities of Hengrave* (London, 1822); Plan of Hengrave Park, 1588, WSRO P746/1.
62. 1613 map by Samuel Piers at Melford Hall.
63. D. MacCulloch, *Suffolk and the Tudors* (Clarendon: Oxford, 1986), p. 115.
64. John Gage, *History and Antiquities of Hengrave* (London, 1822), p. 180.
65. A. G. Dickens (ed.), *The Register or Chronicle of Butleigh Priory, Suffolk* (Winchester, 1931) p. 50.

Notes to Chapter 3. The development of landscape design, *c.* 1660–1735

1. E. B. MacDougall and F. Hamilton Hazelhurst (eds), *The French Formal Garden* (Third Dumbarton Oaks Colloquium on the History of Landscape Architecture, Washington, 1979); R. G. Saisselin, 'The French Garden in the Eighteenth Century: from Belle Nature to the Landscape of Time', *Journal of Garden History* 5, 3 (1985), pp. 284–7.
2. F. Hopper, 'The Dutch Classical Garden and Andre Mollet', *Journal of Garden History* 2, 1 (1982), pp. 25–40.
3. Tom Williamson, *Polite Landscapes: Gardens and Society in Eighteenth-Century England* (Alan Sutton: Stroud, 1995), pp. 24–5.
4. T. Turner, *English Garden Design: History and Styles Since 1660* (Antique Collector's Club, Woodbridge, 1985); M. Hadfield, *A History of British Gardening* (Penguin: Harmondsworth, 1985), pp. 106–78; D. Jacques and A. Jane Van Der Horst, *The Gardens of William and Mary* (Christopher Helm: London, 1988).
5. J. Knyff and J. Kipp, *Britannia Illustrata* (1707; reprinted for the National Trust, London, 1984, edited by John Harris and G. Jackson-Stopps).
6. Letter of attorney 20 January 1666: WSRO HA 513/33/11. This and subsequent sections on the Euston landscape are largely based on the work of Sally Wilkinson: 'Euston: a History of the Landscape Park and Garden', unpublished undergraduate dissertation, School of History, University of East Anglia 1999; and D. Dymond, *The Historical Evolution of Euston Park*, unpublished manuscript, 1995.
7. M. A. Green (ed.), *Calandar of State Papers Domestic Series of the Reign of Charles II, 1666–67* (London, 1864), p. 546.
8. C. Hood, 'An East Anglian Contemporary of Pepys: Phillip Skippon of Foulsham, 1641–1692', *Norfolk Archaeology* 22 (1926), pp. 147–89.
9. E. S. de Beer (ed.), *The Diary of John Evelyn* (Oxford University Press, 1955), vol. 3, p. 591; vol. 4, p. 117.
10. de Beer (ed.), *The Diary of John Evelyn*, vol. 4, p. 117.

11. Easton Tithe Award, ESRO P461/88. Few if any original trees survive, the majority of the timber – oaks – having been felled in the 1960s.
12. Lancelot Brown's plan for Heveningham, Vanneck Papers, Cambridge University Library.
13. ESRO HB 10: 50/20/41.
14. Williamson, *Polite Landscapes*, pp. 36–7, 45.
15. de Beer (ed.), *The Diary of John Evelyn*, vol. 4, p. 117.
16. Coldham Hall Lawshall Tithe Award map, 1842, WSRO T33/1,2; Stanningfield Tithe Award map, 1840, WSRO T137/1,2. Thornham: estate map, 1777, ESRO HD 917/11.
17. J. G. Gazley, *The Life of Arthur Young* (American Philosophical Society: Philadelphia, 1973), p. 1.
18. Northamptonshire Record Office, 127/43.
19. ESRO HA 167 3053/180.
20. Although no map of the site earlier than the Tithe Award map for Lawshall and Stanningfield exists, this arrangement is likely to be an early one: Lawshall Tithe Award map, 1842, WSRO T33/1,2; Stanningfield Tithe Award map, 1840, WSRO T137/1,2.
21. Survey of lands, WSRO E7/10/26; Rushbrooke Tithe Award, WSRO T59/2.
22. Survey of the Manor of Great Saxham, 1729: ESRO T4/33/1.24.
23. Survey of estates belonging to the Earl of Ashburnham and Charles Bone, 1772: HA 1/HB4/1.
24. Also shown on a map of 1741: ESRO P638.
25. 'Dalham Hall, Suffolk', *Country Life* 54 (1923), pp. 280–5.
26. WSRO 279/5.
27. In private ownership: reproduced in *Country Life* 54 (1923), p. 282.
28. Certainly by 1808; they are not shown on a 'Plan of the Park at Dalham … the Property of Sir James Affleck Bart. 1808': WSRO 279/2.
29. ESRO HD417/14.
30. 'Campsea Ash', *Country Life* 18 (1905), pp. 54–62; Kenworthy Brown, *et al.*, p. 221.
31. ESRO HA H/C9/8.
32. ESRO P461/55.
33. The extensive web of avenues postulated by P. F. Springlett must be treated with a measure of scepticism: 'Ashe High House', *Garden History* 2, 3 (1974), pp. 77–89; 'Campsea Ash', *Garden History* 3, 3 (1975), pp. 62–75.
34. ESRO SC 088/4.
35. J. W. Lowther (First Viscount Ullswater), 'The Gardens at Campsea Ash', *Journal of the Royal Horticultural Society* 53, 11 (1922).
36. ESRO SC 088/4.
37. Lowther, *op. cit.*
38. Pevsner, *Suffolk*, p. 413; English Heritage Listed Buildings information.
39. See the two-volume diary of of John Hervey, First Earl, in WSRO: Ac941/13–14.
40. Deed, 1698: WSRO FL557/3/4–5.
41. R. North, *A Discourse of Fish and Fish Ponds* (London, 1713), p. 27.
42. North, *Discourse*, p. 21.
43. 27 June 1798, p. 2, col. 5.

44. 'Survey of the Farm Called Hargate House'; original lost, reproduced in J. Gage, *History and Antiquities of Suffolk* (London, 1838).

45. ESRO HA 176 3050/130

46. Boxford Glebe Terrier 1723: WSRO 806/1/17. It is interesting that the nineteenth-century terriers refer, not to a canal, but to the 'parsonage moat'.

47. Survey of 1730 by William Warren: WSRO M550/1.

48. ESRO S1/2/30.2.

49. E. Martin, T. Easton, and I. MacKechnie, 'Conspicuous Display: the Extra-ordinary Garden and Buildings of a Minor Gentry Family in mid Suffolk', *Proceedings of the Suffolk Institute of History and Archaeology* 37 (1996), pp. 409–27.

50. *Ibid.*, 'Conspicuous Display', p. 73.

51. WSRO M554.

52. ESRO ND 1000/4.

53. E. Martin and A. Oswald, 'The House and Gardens of Combs Hall near Stow-market: a Survey by the Royal Commission on the Historical Monuments of England', *Proceedings of the Suffolk Institute of Archaeology and History* 37 (1996), pp. 409–27.

54. ESRO P638.

55. ESRO HB 8/50/1/62.3.

56. RCHME, National Monuments Record, Swindon, G1174.

57. RCHME, Ntional Monuments Record, Swindon: BR 87/9599 and 9600.

58. WSRO HA 540/3/12.

59. Katherine Doughty, *The Betts of Wortham in Suffolk 1480–1905* (Bodley Head: London 1912), p. 121.

60. This information is from P. Murrel's typescript in the WSRO: 'Sir Dudley Cullum, 3rd Bart of Hawstead Place'. The manuscript reference is E2/18 ofs. 120–121.

61. *Ibid.*, WSRO E2/18 fo. 101, 104).

62. Again, information from Murrell, *op. cit.*

63. Little Thurlow: WSRO HA 540/3/12; Dalham, *Country Life* 54 (1923), pp. 280–5; Hengrave, WSRO 712/58.

64. J. Cullum, *History and Antiquities of Hawsted in the County of Suffolk* (second edn, with additions by T. Cullum, London, 1813), p. 140.

65. Charles Bunbury (ed.), *Memoir and Literary Remains of Lieut-General Sir Henry Edward Bunbury, Bart* (Spottiswoode and Co.: London, 1868), p. 242.

66. Arlington received a licence to empark in 1677; some authorities have suggested that a park may have been laid out before this date but, as Sally Wilkinson has recently demonstrated, there is little real evidence for this.

67. de Beer (ed.), *The Diary of John Evelyn*, vol. 4, p. 117.

68. None of these are listed in the Suffolk *Chorography* of 1602, and while they may have been created before 1660 there is no good evidence for this. Livermere Hall is shown, apparently surrounded by a park, in a painting by Peter Tillemans, dated by Harris to c. 1720: John Harris, *The Artist and the Country House* (Sotheby Parke Bernet, London, 1979), pp. 236–7. Nineteenth-century writers like Shoberl (*Suffolk* (London, 1818), p. 26) asserted that the park was laid out by Baptist Lee, who purchased the estate from the Duke of Grafton in 1722, and

it is certainly true that Lee made a number of minor land purchases in the 1730s (ESRO HA 93 2/706–718), but this may relate to the subsequent expansion of the park. Little Glemham is shown on a map of 1720, and was almost certainly created when the hall was extensively reconstructed a few years earlier. For Ickworth, see the two volumes of diary/expense books of John Hervey, First Earl of Bristol: WSRO Ac. 941 83/1. The making of the park is detailed in the Diary and Expenses of John Harvey, First Earl'; 2 vols, WSRO, Ac. 941 46/13–14. It was largely completed by 1702, when an appendix was added to an existing field book of 1665, entitled 'The Admeasurement of Ickworth Park Pale'. Other evidence for the inception of the park is provided by, for example, an agreement with commoners regarding rights in one of Ickworth's common fields, taken into the park in November 1702: WSRO HA 507/2.

69. Survey of lands, 1734; WSRO E7/10/26.
70. *Ibid.*
71. WSRO HD 1969/1.
72. WSRO HA 540/2/43, lease of 1706.
73. ESRO HA 93/12/78.
74. de Beer (ed.), *The Diary of John Evelyn*, vol. 4, p. 118.
75. See Sally Wilkinson, *Euston: a History of the Landscape Park and Garden*, pp. 6–7.
76. Edward Martin, 'Deserted, Dispersed and Small Settlements', in David Dymond and Edward Martin (eds), *An Historical Atlas of Suffolk* (revised and enlarged edition, 1999), pp. 88–9.
77. ESRO HB 10: 50/20/41.
78. ESRO HA 11 C9/19/2.

Notes to Chapter 4. The triumph of the park

1. C. Hussey, *English Gardens and Landscapes* (Country Life: London, 1967); D. Jacques, *Georgian Garden: the Reign of Nature* (Thames and Hudson: London, 1983).
2. H. Walpole, *The History of the Modern Taste in Gardening* (London, 1770; Gardland edn, New York, 1982).
3. J. Dixon Hunt, *William Kent, Landscape Gardener* (Zwemmer: London, 1987); Kimerley Rorschach, *The Early Georgian Landscape Garden* (Paul Mellon Centre, New Haven, Connecticut: 1983), pp. 50–9. For the suggestion that gardens of this kind often formed components in more geometric schemes see: J. Phibbs, 'Pleasure Grounds in Sweden and their English Models', *Garden History* 21, 1 (1983), pp. 60–90; and Williamson, *Polite Landscapes*, pp. 69–71.
4. Sally Wilkinson: 'Euston'.
5. Historic Manuscripts Commission, *Fifteenth Report, Appendix, Part VI: The Manuscripts of the Earl of Carlisle Preserved at Castle Howard* (London, 1907), pp. 143–4.
6. *Ibid.*, p. 87.
7. C. Hussey, *English Gardens and Landscapes*, p. 156.
8. W. S. Lewis and R. S. Brown (eds), *Horace Walpole's Correspondence*, vol. XVIII (London, 1955), pp. 254–5.
9. Wilkinson, *Euston*, p. 34.

10. Deed, 1698: WSRO FL557/3/4–5.

11. Plan of Culford, T. Wright, 1742: WSRO E8/1/10.

12. Notebook of Thomas Wright, Avery Library Collection, Columbia University, New York, USA.

13. Quoted in G. Storey, 'Culford Hall near Bury St Edmunds'. In *People and Places: an East Anglian Miscellany* (Terrence Dalton: Lavenham, 1973), pp. 94–200; p. 100.

14. Private collection.

15. ESRO 93 2/730.

16. ESRO 93 2/ 731.

17. The earliest maps are undated: a late eighteenth-century map showing the serpentine river in Livermere Park: ESRO HA 93/12/52; and 'A survey of the piece of Water in Livermere park showing the Island', July 1763: HA 93/12/51.

18. D. Stroud, *Capability Brown* (Country Life Books: London, 1965); R. Turner, *Cpability Brown and the Eighteenth-Century English Landscape* (Weidenfeld and Nicholson: London, 1985); R. Williams, 'Making Places: Garden Mastery and English Brown', *Journal of Garden History* 3, 4 (1983), pp. 382–5.

19. Stroud, *Brown*, p. 224.

20. My thanks to David Brown for this information.

21. Stroud, *Capability Brown*, drawing on the evidence of Brown's account book in the library of the Royal Horticultural Society.

22. S. H. A. Hervey (ed.), *Journals of the Hon Wm Hervey in North America and Europe, 1755–1814, with Memoir and Notes* (Bury St Edmunds, 1906), p. 232.

23. Stroud, *Brown*, p. 114.

24. Branches estate, 1766: WSRO T44/4/10.

25. Stroud, *Brown*, pp. 203–5.

26. *Ibid.*

27. My thanks to David Brown for this information. See also Wilkinson, *Euston*, p. 46.

28. Map of Lands in Euston, Fakenham and Rymer, 1772, surveyed by Mr Brown and J. Parker, Euston Estate Office.

29. Stroud, *Brown*, p. 236.

30. Although Brown may have supplied a design for a proposed new house: *ibid.*

31. J. Phibbs/Debois Landscape Survey Group, *Ickworth: a Survey of the Landscape, Part 1: History and Proposals*, unpublished report for the National Trust (East Anglian Region), 1991.

32. Pevsner, *Suffolk*, pp. 246–8; *Country Life* 58 (1925), pp. 432–40, 472–9, 505–15.

33. Lancelot Brown's plans for Heveningham, Vanneck papers, Cambridge University Library.

34. David Lambert, unpublished report on the landscape at Heveningham prior to restoration, n.d.

35. Vanneck was still intending to carry out the plan in 1784, for the French visitor Rochefoucauld, who visited with Lazowski in that year, reported that 'Soon, in the bottom of the valley, there will be a superb artificial river that should have (and I believe it will) a perfectly natural look': N. Scarfe (ed.), *A Frenchman's Year in Suffolk: French Impressions of Suffolk life in 1784* (Suffolk Record Society,

vol. 30, Boydell Press, 1988), p. 141. But nothing ever came of the plan. Perhaps the project lost momentum after Brown's death in 1783.

36. S. C. Roberts (ed.), *A Frenchman in England, 1784: Being the 'Melanges sur le Angleterre' of Francois de la Rochefoucauld* (Cambridge University Press, 1933).

37. *Ibid.*

38. Abstract of title of Sir Thomas Gage's lands, 1779: WSRO 970/4.

39. Map of Fornham St Genevieve 1769: WSRO 373/23; Hodskinson' map of Suffolk, published 1783 but surveyed several years before.

40. Stroud, *Brown*, p. 197.

41. Road order and map, 1787; WSRO Q/SH 34; map of lands of B. Howard, 1788: WSRO 373/24.

42. See F. Cowell, 'Richard Woods (?1716–93). A Preliminary Account': part 1, *Garden History* 14, 2 (1986), pp. 85–120; part 2, *Garden History* 15, 1 (1987), pp. 19–54; part 3, *Garden History* 15, 2 (1987), pp. 115–35.

43. S. Tymms, 'Hengrave Hall', *Proceedings of the Suffolk Institute of History and Archaeology* 1 (1848/53), pp. 331–9; 'Hengrave Hall', *Country Life* 27 (1910), pp. 558–67.

44. Map of 'Lands of Sir Wiiliam Gage', 1742, WSRO 712/58; 'Lands of Sir Thomas Gage Bart., 1769, WSRO 712/60.

45. The original plan is lost. There is a photographic copy in the collection of the Royal Institute of British Architects in London, which is reproduced in Cowell, *op. cit.*, part II, p. 40.

46. Map of Hengrave estate, 1816: WSRO 712/61.

47. Information from David Brown, currently completing his PhD thesis on Richmond at the University of East Anglia.

48. ESRO 5.1/10/49; Pevsner, *Suffolk*, p. 469; Kenworthy-Brown *et al*, pp. 269–70.

49. Humphry Repton to Norton Nichols (of Costessey in Norfolk): 26 August 1788: Bristol University Library 180/1.

50. Norfolk Record Office, MS10 T131B.

51. The suggestion is David Brown's.

52. N. Scarfe, *A Frenchman's Year*, p. 127.

53. Woolverstone Tithe Award, ESRO FDA 298/A1/1b.

54. Survey of Branches estate, 1766: WSRO T44/4/10.

55. ESRO HD 117/31.

56. At Brettenham, for example, his representation suggests a park of around 200 acres, although the Tithe Award map shows a park of only 130 acres and Davy in 1827 describes to the park as 'not large but well laid out' (Brettenham Tithe Award map, 1844: WSRO T 23/1,2; H. Davy, *Views of Seats in Suffolk* (London, 1827, not paginated). In part, such discrepancies really reflect problems of definition: successive sources often appear to show that the area of small parks, in particular, fluctuated. Denston Park, for example, appears to cover around 40 hectares on Hodskinson's map but had significantly 'expanded' to the east by the time the Ordnance Survey draft drawings were made in *c.* 1817, to around 40 hectares. It had contracted again by the time the First Edition 1″ map was surveyed in 1844.

57. Survey of Hintlesham, 1721: ESRO HA 167 3050/130; deed of 1747, ESRO HA 100/A1/51(2).

58. HA 53 359/618: this was shortly after the estate was bought and the house rebuilt by Miles Barnes in 1744. Kirby shows a park here in 1736, but this must have been defunct by the time of purchase, for none is mentioned in the sales documents. See below, pp. 00.

59. ESRO HA 119 50/3.

60. Various sources have attributed the new house to his son, Joshua, on the not unreasonable grounds that Joshua II was 70 years old in 1760; his will, however, specifically refers to the 'mansion house I have lately erected'; typescript history of the Joshua Grigby family, R. L. Workman (1986), WSRO HD1339, p. 67.

61. *Ibid.*, p. 77.

62. The park did not apparently exist at this time – the act mentions only the mansion house, dovehouse and avenues of Thomas le Blanc: WSRO E3/18/5. It had, however, come into existence by 1783, when it is shown on Hodskinson's map. It is first shown in any detail on a map drawn up in connection with a subsequent enclosure, in 1802: WSRO Q/Ri 10.

63. As noted above (pp. 00), a document of 1753 strongly implies that the park had just come into existence; why else would Baptist Lee and James Calthorpe agree to divide their lands by a ha ha unless Ampton park had come into existence to the west of this line: ESRO 93 2/730.

64. Rushbrooke Tithe Award map, 1843, WSRO T59/2; Scarfe (ed.), *Frenchman's Year*, p. 105.

65. Helmingham Hall archives, B1/28/26 and B1/46/10.

66. Daniel Defoe, *A Tour through the Whole Island of Great Britain* (1724; Penguin edition, 1971), pp. 73–4.

67. *Ibid.*, p. 71.

68. *The Gardener's Chronicle*, 1867, p. 157.

69. Map of Nacton Parish, 1805: ESRO HE 8/2652/3.

70. S. Wade Martins and T. Williamson, *Roots of Change*, pp. 12–13, 21–7.

71. Plan of Hengrave Park, 1588: WSRO P746/1: Map of 'Lands of Sir Wiiliam Gage', 1742, WSRO 712/58.

72. Shrubland Hall archives.

73. S. Daniels, 'The Iconography of Woodland in Later Eighteenth-Century England', in D. Cosgrove and S. Daniels (eds), *The Iconography of Landscape* (Cambridge University Press, 1988), pp. 51–72.

74. Red Book for Henham, private collection.

75. Survey of the Manor of Great Saxham, 1729: ESRO T4/33/1.24.

76. Williamson, *Polite Landscapes*, pp. 87–93; M. Laird, *The Flowering of the Landscape Garden* (University of Pennsylvania Press: Philadelphia, 1999).

77. WSRO HD 1155/1.

78. J. Croker (ed.), *Letters of Mary Lepel, Lady Hervey, with a Memoir and Illustrative Notes* (John Murray: London, 1821), p. 106.

79. ESRO HD 331/1 and 332/1.

80. Map of estate of Rt Hon Charles Lord Matnard, 1757, by T. Skynner: ESRO HB 21 280/2.

81. ESRO HA 116/8004.

82. Walpole, *History of Modern Taste*, p. 272.

83. Estate of John Sherman, Melton, 1765: ESRO 51/1/16.

84. *Bury and Norwich Post*, 22 July 1801, p. 1, col. 3.

85. WSRO 317/0.

86. Scarfe (ed.), *Frenchman's Year*, pp. 34–5.

87. J. Phibbs, Debois Landscape Survey Group, *Ickworth Park*

88. See Norman Scarfe, 'Great Saxham Hall', *Proceedings of the Suffolk Institute of History and Archaeology* 26, p. 230. Julia Abel Smith: 'Great Saxham Hall, Suffolk', *Country Life* 27 (1986).

89. Sale by Mure to Mills, WSRO 317/1; Stonemason's wage book, 1796–8, WSRO 2285.

90. Smith, 'Great Saxham Hall', p. 230.

91. Scarfe (ed.), *Frenchman's Year*, pp. 102–3.

92. Field Book of the Estates of Thomas Mills in Great Saxham: ESRO HD 115/2.6.

93. Or so it is argued by Smith, 'Great Saxham Hall, Suffolk'. There are, however, grounds for believing that while this elegant structure clearly does date from the suggested period, it may have been brought from elsewhere and erected on the site in the middle decades of the century. Not only does it not figure in the 1801 survey, it does not appear on the Great Saxham Tithe Award map of 1841, although this shows the other garden buildings: WSRO T51/2.

94. E. D. H. Tollemache, *The Tollemaches of Helmingham and Ham* (Cowell: London, 1949), p. 34.

95. Estate map of Helmingham and surrounding parishes, 1803: ESRO HD 11/475.

96. E. D. H. Tollemache, *The Tollemaches*, p. 120.

97. J. Blatchley (ed.), *David Elisha Davy: A Journal of Excursions through the County of Suffolk 1823–1844* (Suffolk Record Society, vol. 24, Boydell Press: Woodbridge, 1982), pp. 51–2.

98. Scarfe (ed.), *Frenchman's Year*, p. 126.

99. Scarfe (ed.), *Frenchman's Year*, p. 141.

100. It is clearly shown on the Easton Tithe Award map: ESRO P461/88.

101. E. J. Climenson (ed.), *Passgaes from the Diaries of Mrs Phillip Lybbe Powys 1756–1808* (London, 1899), p. 175.

102. ESRO HA 53/359/130.

103. ESRO HA 53/359/130.

104. ESRO HA 53/359/114.

105. ESRO FDA 222/A1/1.

106. Sotterley estate office.

107. WSRO Ac941/83/1. See also J. Phibbs/Debois Landscape Survey Group, *Ickworth: a Survey of the Landscape, Part 1: History and Proposals*, unpublished Report for the National Trust (East Anglian Region), 1991.

108. *Ibid.*

109. WSRO P551.

110. Hervey (ed.), *Journals of the Hon Wm Hervey*, p. 377.

111. Shown on a map of 1785: WSRO M550/3.

112. Survey of the Estate of John Robinson Esq, 1778: WSRO 279/6, MR59). Map

of an Estate of Reginald Rabett Esq., Bramfield and Thorington (1745): ESRO HD 42/1/332/1.

113. Plan of the Hardwick House Estate, 1810; WSRO E2/12/4.

114. WSRO E3/30/13.9.

115. 'A Map of the Estate of the Rt Hon Lady Dowager Chedworth Lying in the Parish of Erwarton': ESRO HE3: 424.

116. Sale particulars of mansion and grounds in Lakenheath, the property of Barnes Caldecott Esq.: ESRO HA 536/CD 33.

117. WSRO E3/36/1.

118. Eleazer Davy closed a footpath leading across his land to the church in 1777, and planted trees around it; the park expanded to the east in 1785 following the construction of the turnpike road: see L. D. Parr, *Yoxford Yesterday*, unpublished typescript ESRO; Tithe Award, Yoxford, 1839, ESRO FDA 305/A1/1b.

119. Benacre estate map, 1778, ESRO T 631 (rolls).

120. ESRO HB10 50/20/41: compare the map of 1738, discussed above, pp. 00.

121. J. C. Loudon, *The Suburban Gardener and Villa Companion* (London, 1838), p. 162.

122. K. D. M. Snell, *Annals of the Labouring Poor: Social change and Agrarian England 1660–1900* (Cambridge University Press, 1983); P. Langford, *A Polite and Commercial People: England 1727–1783* (Oxford University Press, 1992), especially pp. 435–59.

123. N. MacKendrick, J. Brewer, and J. H. Plumb, *The Birth of a Consumer Society* (Europa: London, 19822); Williamson, *Polite Landscapes*, pp. 100–18.

124. Scarfe (ed.), *Frenchman's Year*, p. 128.

125. Rev. R. Canning, 'Introduction' to 1764 edn of J. Kirby's *The Suffolk Traveller*.

126. ESRO HA 11/E1/x.

127. Scarfe (ed.), *Frenchman's Year*, p. 135.

128. *Bury and Norwich Post*, 23 May 1798. p. 2, col. 5.

129. The estate records contain an excellent series of timber and wood accounts from the early nineteenth century: WSRO HA 507/3/494–8; HA 507/3/500.

130. Robert Halsband, *Lord Hervey: Eighteenth-Century Courtier* (Oxford University Press, 1973), p. 132.

131. J. Lawrence, *The Modern Land Stewart* (London, 1801), p. 100.

132. *Bury and Norwich Post*, 6 October 1802, p. 2, col. 5.

133. A survey of 1825 refers to 'the sunk fence or old road': ESRO 475/999.

134. Culford Red Book, private collection.

135. Scarfe (ed.), *Frenchman's Year*, p. 41.

136. WSRO E3/10/53: the area is marked as 'Plantation and Tea House'.

137. Its date is unknown; it appears on an estate map of 1785 (ESRO HA 93/12/59) and was probably constructed in the 1770s, around the time that the new Shrubland Hall was erected, some way away from the old site, a little further along the same escarpment.

138. R. L. Workman, Typescript history of Joshua Grigby Family (1986), WSRO HD1339, p. 79.

Notes to Chapter 5. Repton and the picturesque

1. Williamson, *Polite Landscapes*, p. 141.
2. R. P. Knight, *The Landscape: a Didactic Poem, in Three Books, Addressed to Uvedale Price Esq.* (London, 1794); U. Price, *An Essay on the Picturesque, as Compared with the Sublime and the Beautiful* (London, 1794).
3. W. Gilpin, *Three Essays on Picturesque Beauty, on Picturesque Travel, and on Sketching Landscape* (London, 1792).
4. Turner, *English Garden Design*, p. 124.
5. J. Phibbs, 'The Picturesque Movement', *Architectural Association Conservation Newsletter* 3 (1992), pp. 3–5.
6. Price, *Essay*, vol. I, p. 24.
7. K. Laurie, 'First Years', in G. Carter, P. Goode and K. Laurie (eds), *Humphry Repton, Landscape Gardener* (University of East Anglia, Norwich, 1982), pp. 5–9.
8. The best account of Repton and his work is: S. Daniels, *Humphry Repton: Landscape Gardening and the Geography of Georgian England* (Yale University Press: London, 1999). G. Carter, P. Goode and K. Laurie (eds), *op. cit.*, pp. 5–9.
9. William Mason to William Gilpin, 26 December 1794. Bodleian Library MS (Eng, misc. d. 571. f224).
10. Norfolk Record Office, MS10 T131 B.
11. Humphry Repton, Red Book for Broke Hall, Nacton, February 1792.
12. Red Book for Shrubland Hall, 1789, Shrubland Hall archives.
13. This house was built by John Bacon in the 1770s, to designs by James Paine; it is first shown on a map of 1785: ESRO HA 93/12/59, standing within a small landscape park created by throwing together (and thinning the trees within) the deer park and warren, shown on a map of 1688: ESRO HA 93/12/78. As already discussed, a prospect tower stood some way to the north of the new site. It, too, was positioned on the edge of a steep escarpment, looking over the Gipping valley.
14. ESRO P461/66 and ESRO P461/11. Drawings in Shrubland Hall archives.
15. Red Book for Tendring Hall, 1790: private collection.
16. Red Book for Livermere Hall, 1791, Shrubland Hall archives.
17. There is a mystery here. There had been islands in the lake in the 1760s: they are shown on a map of 1763 (WSRO HA 93/12/51).
18. Red Book for Tatton Park, Cheshire: quoted in P. Goode, 'The Picturesque Controversy', in G. Carter *et al.* (eds), *Humphry Repton*, p. 34.
19. Red Book, Henham Hall, 1791, private collection: John Popham, 'Henham Estate: Report on Site for Proposed Hall in Henham Park', unpublished type-script, 1996.
20. *Ibid.*
21. *Humphry Repton: the Red Books for Brandsbury and Glemham Hall*, facsimile with an introduction by Stephen Daniels (Dumbarton Oaks: Washington, 1994).
22. ESRO HD 1000/2 and HD 11 475/329b.
23. Red Book for Culford, 1792, private collection.
24. WSRO Q/SH24. The 1817 date is provided by the draft drawings of the Ordnance Survey, British Library.

25. Culford estate papers, WSRO HA 513/29.

26. *Bury and Norwich Post*, 9 December 1795, p. 2, col. 4.

27. Red Book for Broke Hall, Nacton, 1792, Shrubland Hall archives.

28. An estate map of 1768 shows the house surrounded by hedged fields: ESRO HA 93/12/38. A small park is, however, showsn here on Hodskinson's county map of 1783. It was probably laid out when the hall was remodelled in the mid 1770s.

29. They are shown on a rough map dated 1803, possibly in the process of being implemented: ESRO HD 11/475/190.

30. ESRO 276/218.

31. Price, *Essay*, III, p. 217.

32. Price, *Essay*, III, p. 217.

33. Williamson, *Polite Landscapes*, pp. 153–4; T. Williamson, *The Archaeology of the Landscape Park* (British Archaeological Reports, Oxford, 1998), pp. 200–2; S. Daniels, 'The Political Landscape', in G. Carter *et al.*, *Humphry Repton*, pp. 110–21.

34. Shrubland Hall archives.

35. Although some of the planting here was added later in the nineteenth century.

36. Lewis Kennedy, *Notitiae on Livermere Park, House and Gardens*, 1815; Shrubland Hall archives.

37. *Topographical and Historical Description of the County of Suffolk* (London, 1813), p. 26.

38. Thornham Tithe Award, ESRO P461/257; notebook concerning planting in park, HA 116/3.

39. Davy, *Seats*.

40. No park is shown on Hodskinson's map of 1783, but when the estate was sold in 1811 the property included a park: Shoberl, *Suffolk*, p. 27; according to the Tithe Award maps of Woolpit (1846: WSRO T119/2); Haughley (1845: WSRO P461/121); and Wetherden (1845: WSRO P461/222), the park covered an area of around 120 acres.

41. First shown on a map of 1810; a map of 1813 shows the names of the previous fields, and their boundaries in dotted outline: ESRO HD 11/475/229; ESRO HA 244 c/26/2.

42. The house was erected *c.* 1810, and there is no sign of an earlier park on this site; the sales particulars of 1830 describe it as having been built 'about twenty years since', and give the area of the park as *c.* 120 hectares: ESROHA 408/1536. Both Kirby in 1766 and Hodskinson in 1783 show a park at Benhall, but this was on a different site, around Benhall Lodge.

43. William Cobbett, *Rural Rides* (London, 1830), pp. 225–6.

44. ESRO HA 408/1536.

45. H. Repton, *Fragments on the Theory and Practice of Landscape Gardening* (London, 1816), p. 69.

46. Kenworthy-Brown *et al.*, *Burkes and Saville's Guide*, p. 238; Pevsner, *Suffolk*, p. 216; ESRO HA22 1335/9/1.

47. ESRO HA22 1335/9/1.

48. ESRO 278/114.

49. *Bury and Norwich Post*, 21 July 1797, p. 2, col. 4.

50. ESRO HA34 50/21/6.1 p. 4–5.

51. ESRO HA 222/335/9/1; HA 43 T501/159.

52. A. I. Suckling, *History and Antiquities of the County of Suffolk*, vol. 3, p. 214.

53. A map of 1802 (in private ownership). Tithe Award for Great Redisham and Ringfield, ESRO P 461/205.

54. Tithe Award for Great Redisham and Ringfield, ESRO P461/205; ESRO FDA 200/A/1.

55. WSRO Q/SH 19.

56. ESRO, Survey of Grundisburgh by Isaac Lenny; Grundisburgh Tithe Award, ESRO P461/115.

57. The park is not shown on Hodskinson's map but was in existence by 1815, when a road closure order terminated a footpath running across it: WSRO B/105/2/1.

58. Davy, *Seats*.

59. No park is shown on Hodskinson's map and none is mentioned in Nathaniel Gurdon's will of 1766 (WSRO E3/10/106 18), nor in Philip Gurdon's marriage settlement of 1778 (WSRO 458/3/15); the park first appears on the enclosure map of 1817, WSRO Q/RI 34, and in enlarged form on the Tithe Award of 1842, WSRO P 461.

60. WSRO Q/SH 68.

61. Hadleigh Tithe Award map, WSRO T/127A, 1, 2; Map Book, estates of Sir William Rowley, WSRO S1/13/33.2.

62. NRO MS10 T131B.

63. No park is shown here on an estate map of 1795: HD 11 475/1134; one is shown on a subsequent estate map, of 1829: HD 11 475/1145. The stated areas, of park and woodland, are from the Marlesford Tithe Award map: ESRO FDA 188/C1/1b. It is possible that the park was created around 1817, when a number of road closures were effected in Marlesford, although none apparently within the precise area of the park.

64. The park appears, schematically, on the draft OS surveyors drawings, but more clearly on the Tithe Award map of 1848 (ESRO P461/183); it is possible that the belts are a later addition.

65. Again, no park is shown by Hodskinson, but one does appear on a map of 1818 (WSRO HA 530/3/1). The extended park is shown on an estate map of 1824 (WSRO E/8/1/11.2).

66. Map of 1818, HA 530/3/1. It is possible that the apparent discrepancies between these two maps are simply the result of differences in definition of the kind already discussed (above pp. 00).

67. R. B. Gilston, 'Brandon Park', *Journal of the Forestry Commission* 19 (1948), pp. 20–3.

68. WSRO.

69. Compare the depiction of the park on Hodskinson's map, which clearly shows the parish church standing on its eastern boundary, with the 1817 enclosure map for Polstead (WSRO Q/RI 34) and the Tithe Award map (WSRO T154/1,2).

70. Road closures, 1816 and 1818: WSRO Q/SH 25 and 26; 6. Map of Drinkstone Park, 1818, reproduced in R. L. Workman, typescript history of Joshua Grigsby

family (1986), WSRO HD 1339, p. 76; Hesset Tithe Award map, 1839, WSRO T80/2.

71. Ousden enclosure map, 1812: WSRO Q/R1 33B.

72. WSRO QSH 29.

73. Orwell Park estate records, ESRO HA 119/50/3/1; Nacton and Levington Tithe Award map 1839, ESRO P461/180; Road closure orders 1818 and 1826: ESRO HA 93/12/16 and 17.

74. Davy, *Seats*.

75. ESRO P461/299.

76. LRO 180/1 AR.

77. ESRO HA 244 c/26/2.

78. Davy, *Seats*; Hintelsham Tithe Award Map, 1839, ESRO P 461/131: Road closure orders, 1825 and 1827 (the latter showing that a lodge had been constructed following the changes effected by the previous closure): ESRO 276/142 and 276/149.

79. Benacre Tithe Award map, ESRO FDA 24/A1/1b.

80. Davy, *Seats*.

81. Shown on an estate map of Helmingham and surrounding parishes, 1803: ESRO HD 11/475.

82. WSRO 317/1.

83. *Ibid.*

84. J. Blatchley and J. James, 'The Beeston-Coyte Hortus Botanicus Gippovicensis and its Printed Catalogue', *Proceedings of the Suffolk Institute of Archaeology and History* 39, 3 (1999), pp. 339–52.

85. *Ibid.*

86. Letters to the Mannock family in Bruges, Belgium: ESRO HD566/1140/1 and 2.

87. E. V. Lucas, *The Colvins and Their Friends* (Methuen: London, 1928), p. 3.

88. H. Lyell (ed.), *The Life of Sir Charles J. F. Bunbury, Bart* (John Murray: London, 1906); Charles Bunbury (ed.), *Memoir and Literary Remains of Liet-General Sir Henry Edward Bunbury, Bart* (Spottiswoode & Co.: London, 1868), pp. 112–13.

89. Shoberl, *Suffolk*, p. 26; this may have appeared as early as 1808, as it seems to be shown on Isaac Lenny's map of that date: WSRO 2399/1.

90. WSRO HD 1535/10.

91. *The Gardener's Chronicle*, 1911, 418.

92. WSRO 941/81/1.

93. WSRO 298.

94. Humphry Repton, *Fragments on the Theory and Practice of Landscape Gardening* (London, 1816), p. 69.

95. WSRO HA 535/5/36; HA 535/5/99.

96. The Landscape Partnership, *Nowton Park*, unpublished report for St Edmundsbury Borough Council. Jane Fiske, *The Oakes Diaries* (Suffolk Record Society/ Boydell: Ipswich, 1990).

97. Shown on a map of 1824, WSRO HA 535/5/34. This may have been recently completed, as a road closure order removed a footpath running through the Paddock opposite the Cottage in 1821: WSRO Q/SH 77.

98. WSRO HA 535/5/1.
99. Map of 1827: HA 535/5/34.
100. Map of the Oakes' estate, 1832: WSRO HA 535/5/34. The phrase 'park or paddock' is used on another road closure order, diverting the public road further away from the Cottage: WSRO Q/SH 80.
101. WSRO HA 535/5/3.
102. *Bury and Norwich Post*, 5 April 1809, p. 3, col. 3.

Notes to Chapter 6. Victorian gardens

1. The best account of Victorian gardens is Brent Elliott, *Victorian Gardens* (Batsford: London, 1986).
2. Shrubland Hall archives, S/A 1/1/5.
3. *The Stranger's Illustrated Guide to Bury St Edmunds, Suffolk* (London, 1871); *Bury St Edmunds and District Illustrated* (Bury St Edmunds, 1906); *Guide to Bury St Edmunds* (Cheltenham, 1907).
4. Elliott, *Victorian Gardens*, pp. 87–90, 123–8.
5. *Cottage Gardener*, September 1856, p. 470.
6. C. Hussey, 'Shrubland Park', *Country Life* 114 (1953), pp. 948–51, 1654–7, 1734–8.
7. *The Florist*, May 1856, p. 54. *Cottage Gardener*, October 13, 1857, p. 34.
8. *The Florist*, May 1856, p.154
9. *Ibid.*
10. Shrubland Hall archives, S/A 1/2/30.
11. Elliott, *Victorian Gardens*, pp. 74–8.
12. *Cottage Gardener*, September 1856, p. 469.
13. *The Florist*, May 1856, p. 53.
14. *The Gardener's Chronicle*, September 1867, p. 1123.
15. *Cottage Gardener*, September 1856, p. 471
16. *The Florist*, May 1856, p.154
17. Shrubland Hall archives, SA 1/1/7.
18. Shown on an undated design by A. Roos in the Shrubland Hall archives, uncatalogued.
19. Shrubland Hall archives, SA 10/1.
20. *Cottage Gardener*, 6 October 1854, p. 6.
21. *The Florist*, May 1856, p. 152.
22. *Cottage Gardener*, 13 October 1857, p. 18.
23. *Cottage Gardener*, 23 September 1856, p. 453
24. *Cottage Gardener*, 13 October 1857, p. 19.
25. Shrubland Hall archives, SA 1/1/8.
26. *Ibid.*
27. *Ibid.*
28. *Ibid.*
29. *Cottage Gardener*, 13 October 1857, p. 33.
30. Newspaper cutting in the Shrubland Hall archives.
31. E. Adveno Brooke, *The Gardens of England* (London 1858).
32. *The Gardener's Chronicle*, 2 November 1867, p. 1123.

33. *The Florist*, May 1856, p. 154.
34. Shrubland Hall archives, SA 10/1.
35. Shrubland Hall archives, SA 10/1.
36. Shrubland Hall archives, SA 10/1; all the documents used in the following discussion are in this section of the archive.
37. 'Somerleyton Hall', *Illustrated London News*, 10 June 1857, pp. 24–5.
38. Crowe's *Handbook to Lowestoft* (Crowes: Norwich, 1853), pp. 172–4.
39. *Ibid.*
40. The accounts of 1851 and 1853 describe a garden very different from that depicted on the Somerleyton estate map of 1857 (Somerleyton Hall archives), or as described and illustrated in the Somerleyton Sales particulars of 1861 or 1862: ESRO HA 236/2/165; ESRO FSC/364/3.
41. C. Ridgeway, 'William Andrews Nesfield: between Uvedale Price and Isembard Kingdom Brunel', *Journal of Garden History* 13 (1993), pp. 69–89.
42. ESRO HA 236/2/165.
43. ESRO SC 479/6.
44. *The Gardener's Chronicle*, 1867, pp. 156–8.
45. ESRO HA 11/C46/28.
46. ESRO HA 11/C46/25 and 31.
47. I am extremely grateful to Christopher Ridgeway, archivist at Castle Howard, for this information.
48. Pevsner, *Suffolk*, p. 341.
49. *Gardener's Chronicle*, 1881, vol. 1, p. 723; 1912, p. 5.
50. Hengrave Tithe Award map, 1839: WSRO 449/3/15.
51. Cambridge University Library, Hengrave Collection; I am most grateful to Edward Martin for drawing this material to my attention.
52. Hengrave and Coldham, Bills, 1866: WSRO 712/95/1–9.
53. ESRO S1/2/300, 68.
54. ESRO HA 61: 436/1,4; HA 93/3/250, 251, 252.
55. *The Gardener's Chronicle*, 1876, pp. 198–9, 205, 229–30.
56. WSRO E2/22/4.
57. *The Gardener's Chronicle*, August 1885, p. 250; *The Gardener's Chronicle*, VI, no. 140 (1889), p. 240.
58. Journal of a Visit to Hardwick, 1869; WSRO 1888.
59. Edward Farrer, *Hardwick Manor House*, 1928.
60. Examples include Bradfield Hall, created when the house was rebuilt (with gardens by Nesfield constructed in 1857); and Nowton Court, created by the expansion of the small pleasure grounds described above when the house was rebuilt in 1875: below, pp. 00.
61. The additon carries a datestone.
62. There are a number of architectural drawings, currently uncatalogued, in the archives at Shrubland Hall.
63. *The Gardener's Chronicle*, 1867, pp. 229–30.
64. Map of 1795, entitled 'The Tower Farm in Freston, the property of Charles Berners': ESRO HD 11 475/1819.
65. ESRO Gc 15: 52/6/19.4.

66. ESRO HA 236/2/165–166.
67. ESRO HA 236/2/239.
68. ESRO HA 236/2/165–166.
69. *The Florist*, 1867, pp. 328–9.
70. Sale particulars, 1836: ESRO HA 408/1541; Tithe Award map, 1847, ESRO P461/25.
71. *The Gardener's Chronicle*, 1876, vol. II, p. 198.
72. *The Gardener's Chronicle*, 1883, vol. II, pp. 557–8.
73. *The Gardener's Chronicle*, 1876, vol. II, p. 198.
74. *The Gardener's Chronicle*, 1876, vol. II, p. 199.
75. ESRO HA 119 (562) 96: miscellaneous bundle of documents, Orwell Park.

Notes to Chapter 7. The business of gardening

1. WSRO 317/1, p. 38.
2. ESRO HA 61 436/547; ESRO HD 1873/1.
3. *Suffolk Mercury or Bury Post*, 9 September 1734, p. 53.
4. WSRO HA 540/3/12.
5. J. Harvey, *Early Nurserymen: with reprints of documents and lists* (Phillimore: Chichester, 1974), p. 71.
6. WSRO 2326/3.
7. Harvey, *Early Nurserymen*, p. 99
8. *Bury and Norwich Post*, 13 January 1802, p. 2, col. 5.
9. *Bury and Norwich Post*, 23 December 1801, p. 2, col. 6.
10. WSRO 317/1, p. 61.
11. Hervey, *Journals of the Hon Wm Hervey*, p. 490.
12. pp. 104–5.
13. ESROHA 134/2706/1.
14. p. 150.
15. p. 150.
16. ESRO HA C2/21.
17. WSRO E2/18 fo. 101.
18. *Bury and Norwich Post*, 10 May 1797, p. 3, col. 4.
19. WSRO 941/81/8.
20. Hengrave and Coldham, Bills, 1866: WSRO 712/95/1–9.
21. Garden diaries, Boulge Hall, 1896: ESRO HA 244/x/5/1.
22. *Bury and Norwich Post*, 16 December 1801, p. 3, col. 4.
23. *Bury and Norwich Post*, 24 March 1802, p. 3. col. 5.
24. WSRO E2/22/4.
25. *Bury and Norwich Post*, 25 November 1801, p. 2, col. 4.
26. D. Grace and D. Phillips, *Ransomes of Ipswich: A History of the Firm and Guide to its Records* (Institute of Agricultural History, Reading, 1975), pp. 3–6..
27. *Bury and Norwich Post*, 9 December 1795, p. 2, col. 4.
28. K. Doughty, *The Betts of Wortham in Suffolk 1480–1905* (Bodley Head: London, 1912).
29. ESRO 408/C2/91.

30. Shrubland Hall archives, SA 10/1.
31. Frederick Mullally, *The Silver Salver* (Granada: London, 1981), p. 43.
32. Shrubland Hall archives, SA/10/3.
33. *The Gardener's Chronicle*, 1876, vol. II, p. 230.
34. Shrubland Hall archives, SA 10/1.
35. In private ownership: reproduced in *Country Life* 54 (1923), p. 282.
36. ESRO HA 11/C46/28.
37. Lancelot Brown's plan for Heveningham, Vanneck Papers, Cambridge University Library: Map of Branches estate, 1766: WSRO T44/4/10.
38. H. Repton, *Fragments on the Theory and Practice of Landscape Gardening* (London, 1816), p. 167.
39. J. C. Loudon, *Encyclopaedia of Gardening* (London, 1822), p. 725.
40. Above, pp. 00.
41. Pevsner, *Suffolk*, p. 411. The hall is shown beside the walled garden on an estate map of 1824 by Lenny, WSRO E/8/1/11.2, and on the Tithe Award map, 1843, WSRO T109/1,2, but had been moved to its new site by the time of the First Edition OS 6″, 1886.
42. WSRO T78/2.
43. Red Book for Livermere Hall, 1791, Shrubland Hall archives.
44. Red Book for Henham Hall, 1791, private collection.
45. Hoxne Tithe Award map, ESRO HD40/422.
46. *The Gardener's Chronicle*, 1876, vol. II, pp. 229–30.
47. M. Williams, *A Study of Kitchen Gardens in Suffolk* (unpublished MA Dissertation, University of East Anglia, 1999), pp. 4–10.
48. Loudon, *Encyclopaedia*, p. 732.
49. Williams, *Kitchen Gardens*, p. 4.
50. *Ibid.*
51. O. R. Wellbanke, *Suffolk, My County* (Heath Cranton, 1934), p. 9.
52. WSRO HA 540/3/12.
53. Murrel, *op. cit.*
54. HA 530/2/34–37 256D. B4.
55. ESRO HD 80/1/1/36–44.
56. J. C. Loudon, *The Suburban Gardener and Villa Companion* (London, 1838), p. 108.
57. WSRO HA 520/2/34–7, 256D. B4.
58. ESRO SC 088/4.
59. Crowe's *Handbook to Lowestoft* (Crowes: Norwich, 1853), pp.
60. ESRO HA 236/2/165.
61. ESRO HA 244 x/5/1.
62. Shrubland Hall archives, SA 10/3.
63. ESRO SC 088/4.
64. J. McCann, *The Dovecotes of Suffolk* (Suffolk Institute of Archaeology and History: Ipswich, 1998), pp. 78–80.
65. *Ibid.*, pp. 81–4.
66. WSRO HA 520/2/34–7, 256D. B4.

Notes to Chapter 8. The late nineteenth century and beyond

1. For the effects of the depression on Suffolk, see Edward Bujak, 'Suffolk Land-owners: an Economic and Social History of the County's Landed Families in the Late Nineteenth and Early Twentieth Centuries' (unpublished PhD thesis, University of East Anglia).
2. William Robinson, *The Wild Garden* (John Murray: London, 1870).
3. William Robinson, *The English Flower Garden* (John Murray: London, 1883).
4. G. Jekyll, *Wood and Garden* (Longmans, Green and Co.: London, 1899). The best account of Jekyll and her style is: Richard Bisgrove, *The Gardens of Gertrude Jekyll* (Frances Lincoln: London, 1992).
5. David Outsell, *The Edwardian Garden* (New Haven and London, 1989), p. 120.
6. *The Gardener's Chronicle*, December 1890, p. 725.
7. Elise Perciful, 'Arts and Crafts Influences in East Anglian Gardens: Gardens and Gardening in Norfolk and Suffolk 1885–1914' (Unpublished PhD Thesis, University of East Anglia, 1999), pp. 209–21.
8. WSRO EF 506/6/1136.
9. Perciful, 'Arts and Crafts Influences'.
10. *The Gardener's Chronicle*, 1888, p. 328.
11. *Ibid.*, 1890, p. 278.
12. *Ibid.*, 1890, p. 795.
13. Shrubland Hall archives, uncatalogued.
14. Shrubland Hall archives, map chest, uncatalogued.
15. Perciful, 'Arts and Crafts Influences', p. 105; photographs in the Jarman Collection, WSRO.
16. Perciful, 'Arts and Crafts Influences', p. 115.
17. WSRO HD 1756/6.
18. Perciful, 'Arts and Crafts Influences', p. 116.
19. *The Gardener's Chronicle*, June 1896, p. 440.
20. ESRO HA 244 D1/H.
21. ESRO HA 244 c/26/3.
22. ESRO HA 244 x/5/1.
23. Perciful, 'Arts and Crafts Influences', p. 119.
24. ESRO HA 79/2/2.
25. Perciful, 'Arts and Crafts Influences', p. 113
26. *The Gardener's Chronicle*, November 1901.
27. Perciful, 'Arts and Crafts Influences', p. 90; *The Gardener's Chronicle*, August 1885, p. 183; October 1896, p. 492.
28. Perciful, 'Arts and Crafts Influences', p. 140; G. Martelli, *The Elveden Enterprise* (London, 1951), p. 55; Shrubland Hall archives, SA 10/3.
29. Sales Particulars, 1895, WSRO FL 574/13/3. Jean Farquar, *Arthur Wakerley, 1862–1931* (London, 1984), p. 5; Sales Particulars, 1918, WSRO OGOD 78.4.
30. Sales particulars in possession of the owners.
31. WSRO E3/36/1.
32. Bawdsey estate map, 1727:ESRO HA 30 50/22/26.1.
33. Bawdsey Tithe Award Map, 1843: ESRO FC 162/ C3 /1.

34. Bawdsey Estate Sale Catalogue 1959: ESRO SC 032/1; Bawdsey Estate Sale Catalogue 1991: ESRO SC 032/8; Guide to Bawdsey Manor, privately published leaflet, n.d., ESRO.

35. *The Gardener's Chronicle*, December 1908, pp. 406–9.

36. *Ibid.*

37. This discussion is based on: Hazel Conway, *People's Parks: The Design and Development of Victorian Parks in Britain* (Cambridge, 1991); and Harriet Jordan, 'Public Parks 1885–1914', *Garden History* 22, 1 (1994), pp. 85–113.

38. Jordan, *op. cit.*, p. 85.

39. This account is largely based on R. Malster, *Lowestoft: East Coast Port* (Terence Dalton: Lavenham, 1982), pp. 53–5.

40. Perciful, 'Arts and Crafts Influences', p. 289

41. At Bury the situation was similar; the Abbey Gardens, although they had been open to subscribers since their inception, only opened to the public on a wider basis from the 1880s. The entrance fee, at one shilling for an adult and sixpence for a child, was high, presumably intentionally so, to exclude undesireable elements and maintain the place as an arena for polite society. Only in 1912 did the Borough Council take a lease on the property, and it was not until the 1930s that the layout of grounds began to change significantly. All in all, Suffolk was tardy in its provision of public parks.

42. Acquisition Report, Downham Hall Estate, Forestry Commission Archives, Great Eastern House, Cambridge.

Index